Regulation of Growth in Neoplasia

# Regulation of Growth in Neoplasia

Editor: *G. V. Sherbet*

Cancer Research Unit, University of Newcastle upon Tyne,
Newcastle upon Tyne, England

16 figures and 7 tables, 1981

S. Karger · Basel · München · Paris · London · New York · Sydney

National Library of Medicine, Cataloging in Publication
    Regulation of growth in neoplasia
    Editor, G.V. Sherbet. – Basel; New York: Karger, 1981
    1. Cell Division   2. Cell Transformation, Neoplastic   3. Cells, Cultured   4. Neoplasms – etiology   I. Sherbet, G.V. (Gajanan V.)
    QZ 202 R344
    ISBN 3–8055–2305–X

All rights reserved.
    No part of this publication may be translated into other languages, reproduced or utilized in any form or by any means, electronic or mechanical, including photocopying, recording, microcopying, or by any information storage and retrieval system, without permission in writing from the publisher.

©   Copyright 1981 by S. Karger AG, P.O. Box, CH-4009 Basel (Switzerland)
    Printed in Switzerland by Werner Druck AG, Basel
    ISBN 3–8055–2305–X

# Contents

Preface .................................................... IX

## Regulation of Growth and Cell Division in the Whole Organism
*John D. Simnett*, Newcastle upon Tyne

| | |
|---|---|
| I. Introduction ................................................. | 1 |
| II. Types of Growth Phenomenon ................................. | 3 |
|     A. Embryonic Growth ......................................... | 3 |
|     B. Developmental Growth ..................................... | 4 |
|     C. Cell Turnover ............................................. | 4 |
|     D. Diurnal Variation in the Rate of Cell Division ................ | 5 |
|     E. Wound Healing, Compensatory Growth and Regeneration ..... | 5 |
|     F. Hypertrophy and Hyperplasia ............................... | 6 |
|     G. Oedema .................................................. | 7 |
|     H. Neoplasia ................................................ | 7 |
|     I. Cell Loss ................................................. | 8 |
| III. The Relative Importance of Cell Programming and of Regulation .. | 8 |
| IV. Common Factors in the Control of Cell Division ................. | 11 |
| V. Tissue Response to Trauma ................................... | 14 |
| VI. Cell Division and Tissue Trauma .............................. | 16 |
| VII. Theories on the Control of Cell Division following Tissue Trauma .. | 17 |
|     A. Wound Hormones ......................................... | 17 |
|     B. Growth Regulation through Negative Feedback .............. | 17 |
|     C. Growth Regulation by Physiological Load .................. | 19 |
|     D. Possible Function of the Lymphoid System in Growth Regulation | 20 |
|     E. Control of Growth by the Nervous System .................. | 21 |
|     F. Biological Factors Affecting the Growth of Cells in Culture .... | 23 |
|     G. Action of Specific Hormones on Cell Division and Growth .... | 23 |
|     H. Growth Control by Tissue Interaction ...................... | 24 |
|     I. Some Common Assumptions Shared by Hypotheses of Growth Regulation | 25 |
| VIII. Regulation of Growth in Lung Alveolar Tissue ................... | 25 |

Contents

IX. The Possible Role of Acute Inflammation and Vascular Changes in the Regulation of Cell Division .................................................. 28
    A. Acute Inflammation ............................................... 28
    B. Changes in Vascular Supply ...................................... 29
X. Control of Blood Flow in the Microcirculation ......................... 30
    A. Anatomy of the Vascular Bed ..................................... 30
    B. Neural Control of Vascular Resistance ............................ 30
    C. Regulation of Blood Flow in Specific Organs ...................... 31
XI. Chemical Mediation of the Proposed Neural and Vascular Effects ........... 32
    A. Blood Flow and the Exchange of Metabolites ...................... 32
    B. Chemical Mediators of Acute Inflammation and Their Effect on Cell Division ........................................................... 33
XII. Evaluation of Growth Control Hypotheses .............................. 34
    A. Wound Hormones ................................................. 35
    B. Negative Feedback and the Chalone System ........................ 35
    C. Physiological Load ................................................ 36
    D. Lymphoid System in Growth Regulation ........................... 36
    E. Control of Growth by the Nervous System ......................... 36
    F. Hormones and Other Biological Factors Stimulating Cell Division .... 37
XIII. Control of Cell Proliferation in the Different Types of Growth Phenomenon . 37
References .................................................................. 39

# Growth Differentiation and Function of Tumours in Organ Culture

*Gisèle M. Hodges*, London

I. Introduction .............................................................. 52
II. Methodology for Tumour Organ Culture ................................ 53
    A. Organ Culture Systems .......................................... 53
    B. Factors Influencing Organ Culture ................................ 55
    C. Analysis of Organ Culture ....................................... 58
III. Characteristics of Tumour Organ Culture ............................... 59
    A. In vitro Longevity: Population Survival and Selection .............. 59
    B. Growth ......................................................... 65
    C. Differentiation and Function ..................................... 69
IV. Experimental Investigations ............................................. 78
    A. Tumour Response to Hormones .................................. 78
    B. Tumour Invasiveness ............................................ 82
    C. Tumour Response to Chemotherapeutic Agents and Ionizing Radiation .. 89
    D. Tumour Response to Virus Infection .............................. 93
V. Concluding Comments ................................................. 95
Appendix .................................................................. 97
References ................................................................ 116

## Control of Proliferation of Normal and Neoplastic Cells in Culture

*P.A. Riley*, London

| | |
|---|---|
| I. Introduction | 131 |
| II. Control of Proliferation | 132 |
| III. Hierarchical Controls | 133 |
| IV. Intrinsic Regulation of Proliferation | 134 |
|     A. Cell Division | 134 |
|     B. Control Point | 137 |
|     C. Inducers of DNA Synthesis | 138 |
|     D. Response to External Factors | 140 |
| V. Extrinsic Factors | 140 |
|     A. Microenvironmental Effects (Intercellular Controls) | 140 |
|     B. Resource Depletion | 158 |
|     C. Metabolic Products | 171 |
| VI. Interpopulation Controls | 178 |
|     A. Hormones | 178 |
| VII. Conclusions | 184 |
|     References | 185 |
| Subject Index | 199 |

# Preface

This book is a sequel to the volume entitled 'Neoplasia and Cell Differentiation' published in 1974. The great interest in the field of Neoplasia and Cell Differentiation has prompted us to bring out this volume, which presents three extensive and authoritative reviews of growth processes and their regulation, especially in neoplasms in vitro. I have little doubt that these will stimulate much thought and research into the mechanisms involved in the control of growth processes in the two closely allied fields of neoplasia and cell differentiation.

I would like to thank the contributors for the assiduity and enthusiasm with which they devoted themselves to the writing of the reviews. I would also like to thank the Publishers for a cool and competent handling of this project which has resulted in a well-produced book.

Newcastle upon Tyne, June 1981 *G.V. Sherbet*

# Regulation of Growth and Cell Division in the Whole Organism

*John D. Simnett*

University of Newcastle upon Tyne, Department of Pathology, Royal Victoria Infirmary, Newcastle upon Tyne, England

## I. Introduction

Current methods of treating cancer involve drastic measures. Surgery is disfiguring while radiotherapy and chemotherapy often make the patient feel extremely weak and debilitated. None of these treatments is particularly effective since, with some notable exceptions, the best that can be hoped for is a few years remission of the illness. Advances that have been made during recent years have not entailed the introduction of new methods so much as the refinement of existing methods, in particular by combining different treatments and by empirical determination of the optimum dosages and means of administration. Over the past decade or more, there has been a growing interest in the immune surveillance theory of cancer, which proposes that the body is capable of recognising and rejecting potentially malignant cells and that cancer develops because of a defective surveillance mechanism. This theory has come under criticism for a number of reasons: in many spontaneous cancers there is no evidence for tumour-associated antigens, in experimental systems the rate of production of T lymphocytes does not correlate well with the incidence of tumours nor is there good evidence that natural killer (NK) cells protect against tumours [110]. Nevertheless, clinical trials involving manipulation of the immune response frequently improve the chances of survival in patients with some forms of leukaemia and carcinoma [201].

A common feature of all the forms of treatment so far mentioned, namely surgery, radiotherapy, immunotherapy and the use of cytotoxic drugs, is that their object is to eradicate the tumour tissue by destroying the cells of which it is composed. The need for such a strategy is based upon the widely held view that in changing from the normal to the neoplastic state, cells have undergone some form of irreversible change. While the main body of clinical observation is consistent with this idea, long-term or even complete remissions, sometimes in the absence of conventional treatment, are by no means uncommon and even when those remissions which seem explicable on the basis of immunological rejection are excluded, there remains a considerable body of evidence, both clinical and experimental, which suggests that the neoplastic state of some tissues can be reversed. Turning to the biological evidence, there is much less support for the view that neoplasia is irreversible. One firmly held belief about neoplasia is that the cells involved have undergone some form of true genetic change or 'somatic mutation'. If this were true, then neoplasia would indeed be irreversible, since the chances of any particular chromosome aberration or gene mutation reverting to normal genotype are very small. While it is true to say that some forms of neoplasia are associated, possibly causally, with a true change in genotype, such evidence is lacking in the majority of cases and in many forms of cancer the biological and clinical data may actually be inconsistent with the somatic mutation hypothesis [209]. There is much evidence to support the idea that cancer is a form of atypical cell differentiation due to changes in the mode of gene expression rather than alteration of genotype and that, in a large proportion of tumours, genetic abnormalities are observed because aberrant cells, which occur even in normal tissues, have a better chance of survival in the neoplastic tissue.

The idea of neoplasia as a state of abnormal differentiation is therefore tenable and this has important implications. While differentiation tends to be a very stable state, it is far from irreversible. Modification of the differentiated state is observed during the modulation of cultured tissues [109], in states of pathological metaplasia or in regenerating tissues and changes in the way in which genes are selectively expressed may be observed following experimental procedures such as nuclear transplantation and cell fusion [209]. Cancer research programmes actively support investigations into normal cell physiology and, in particular, into the processes regulating normal cell growth and differentiation. Implicit in this policy is the idea that it may eventually be possible to control neoplasia by manipulation of the natural physiological controls. This is an exciting, if at present distant, prospect.

It is difficult to find any simple feature which distinguishes tumours from normal tissues. The state of neoplasia rather involves a spectrum of related changes. These include alterations in the biochemistry and morphology of individual cells, invasiveness leading to infiltration of surrounding normal tissues, a loss of cellular adhesion resulting in metastasis and the formation of secondary growths and an imbalance in cell proliferation such that the rate of cell production exceeds the natural loss so that the tumour gradually increases in size. It is this uncontrolled growth which is the most serious clinical feature of the disease. If the physiological control of cancer growth is to become a clinical reality, it must be based upon manipulation of the mechanisms which regulate growth in normal tissues. Growth of the whole organism and of its component organs and tissues is, with few exceptions, due to an increase in the number of component cells and an understanding of how cell division and proliferation is controlled is therefore essential. It is reasonable to ask whether there is a single mechanism which can account for all growth phenomena or whether the means of control differs according to the type of growth process. As an introduction to this enquiry, it is relevant to list the many types of growth in which we might assume some type of cell division control to exist. A detailed review of cell division kinetics and of the methods and terminology employed is given elsewhere [2].

## II. Types of Growth Phenomenon

### A. Embryonic Growth

The first few 'cleavage' divisions of the single-celled zygote are characterised by a short cell-cycle time in which, following mitosis (M), the cell proceeds directly into DNA synthesis (S) with no $G_1$ or $G_2$ stage [101]. During this initial period cell division is synchronous, continuing for at least 6 or 7 consecutive cleavage divisions, but eventually synchrony breaks down with the appearance of the $G_1$ stage and a general lengthening of the cell cycle [190]. This transition is accompanied by the appearance of specific morphogenetic functions which denote the onset of cell differentiation and which probably involve the transcription of new mRNA. In mammals, in which the embryonic cells have a small cytoplasmic mass, the information needed for new morphogenetic functions may be acquired during early cleavage stages [227] while in larger-celled embryos, such as those of Amphibia, the onset of new DNA-RNA transcription appears to occur some time during gastrulation [68]. Some time during the gastrula stages, regional differences appear in the rate of incorporation of specific amino acids into proteins [74] and by the end of gastrulation in Amphibia specific antigenic differences can be demonstrated between neural and epidermal ectoderm, mesoderm and endoderm [60]. The early stages of histotypical differentiation in the adenohypophysis [18] and the kidney tubules [78] are characterised by the appearance of specific antigens and this may be a common feature in all forms of tissue differentiation. Nuclear division, occurring during mitosis, appears to be essential for the transcription of new RNA [100] and the expression of new developmental information [227].

## B. Developmental Growth

This term is used to denote the period of growth extending from the early stages of cell differentiation up to the time when the organism attains its upper size limit. It should be noted that in some species, such as many teleost fish [63], developmental growth is continuous and, given adequate nutrition, no upper size limit is reached. In many animal species developmental growth is at first exponential [35] and in mammals this phase lasts through foetal life into the early post-natal period when the maximum rates of growth may be observerd [220]. Overall exponential growth of an organ may conceal wide differences in the growth rate of its component tissues. Within the intestine, which as a whole presents an exponential growth pattern until the time of birth, the rate of proliferation of epithelium is considerably higher than that of the mesenchyme [202].

Quite early in development, three stages of the cell cycle – S, $G_2$ and M – achieve a fairly constant duration, and thereafter the gradual reduction in cell division rate is due mainly to a lengthening of $G_1$ and a progressive reduction of the growth fraction or proportion of cells actively dividing [35]. This even progression may be interrupted by events such as metamorphosis in insects [250] or puberty in mammals where the sexual [176] and secondary sexual [122] organs may show massive, though temporary, increases in the rates of growth and cell multiplication.

Termination of the period of developmental growth and the achievement of the upper size limit does not mean that growth processes then cease, as will be shown in the following examples.

## C. Cell Turnover

Cessation of overall growth occurs when a state of dynamic equilibrium is reached between production of new cells and the loss of old cells. In many tissues this process of cell turnover is very rapid, for example in parts of the gastrointestinal tract it may be equivalent to total replacement of the epithelial cell population every 2 days [257].

In some tissues, notably the epidermis and intestinal epithelium, cell production is due to the proliferation of 'stem cells' which retain the ability to divide while the maturing and functional cells of the tissue lose the ability to divide and are lost by desquamation [99]. In contrast, the mature liver, kidney [38] and lung [26] do not appear to contain a stem cell population and growth occurs as the result of the proliferation of fully matured functional cells.

It appears that the curtailment and eventual cessation of developmental growth is, in the final stages, due more to an increase in the rate of cell loss than to a reduction in the rate of cell proliferation.

### D. Diurnal Variation in the Rate of Cell Division

In tissues which exhibit some measurable rate of cell turnover, the rate of cell production usually shows a marked and regular pattern of variation throughout the day [77]. During the 24-hour period the difference between the maximum and minimum rates of cell division may, in some organs, be as much as four-fold and the time of maximum and minimum rates appears to differ from organ to organ [212]. When animals or human subjects are kept in conditions where environmental cues, such as variation in light intensity, are absent, the strict 24-hour or diurnal cycle breaks down and the subject enters a state of 'free cycling' where the rate of cell division develops a 'circadian' pattern which has a period sometimes shorter and sometimes longer than 24 h [42]. This suggests that animals have an inbuilt timekeeper, possibly mediated by the hypothalamus, which is programmed to an accurate diurnal cycle by environmental changes.

### E. Wound Healing, Compensatory Growth and Regeneration

Most animals have some means of repairing damage due to physical trauma and tissue loss. In some cases, repair is limited to the restoration of structural integrity in which damaged functional tissue is replaced by connective tissue. This process is termed wound healing by 'second intention' [244]. In other cases full function may be restored by growth due to the proliferation of cells in the remaining mature functional tissue, a phenomenon known as compensatory growth. In some tissues, for example skin and lachrymal gland [98], the response to damage is purely local: increased cell division is restricted to tissues adjacent to the wound.

In other cases, a response may be found in parts of the organ situated at some distance from the site of injury, as in the liver, or in the undamaged contralateral organ, as is the case in the salivary gland, kidney, lung, testis and ovary [99]. In the case of organs such as the liver [147], kidney [10] or lung [214] which exhibit a contralateral response, a local response may also be found where the extent of damage is small.

In many cases of local damage the essential architecture of the tissue must be preserved if perfect restoration is to occur, otherwise the lost functional tissue is replaced by connective tissue. In zonal necrosis of the liver, perfect healing depends on the reticulin framework of the lobule remaining undamaged [244], in skeletal muscle maintenance of intact sarcolemmal tubes [151] and basement membrane architecture is essential [50], while in nerves regrowth occurs only if the connective tissue sheath remains intact [99].

Regeneration of complete body parts such as limbs may sometimes occur. This is frequently regarded as a feature of 'primitive' species but in fact wide differences in regenerative ability may be found between phylogenetically related groups, such as crustaceans and insects (Arthropoda), which suggests that regeneration is a faculty which evolves in cases where it has some adaptive survival value [98]. In many cases, such as the lizard tail [219] or the Amphibian limb [236], regneration depends on the dedifferentiation, proliferation and redifferentiation into new cell types of tissue elements adjacent to the wound. This process may be termed 'physiological metaplasia'.

### F. Hypertrophy and Hyperplasia

Many organs or tissues can become enlarged above their normal size. This may involve enlargement of the individual functional or parenchymal cells (hypertrophy), an increase in the number of such cells (hyperplasia) or both. In the human body, the most extreme form of hypertrophy occurs in the myometrium of the uterus where, during pregnancy, individual smooth muscle fibres increase to ten times their normal resting size [244]. At the time of birth, the average weight of the heart of human infants is 30 g and after this no further cardiac muscle fibres are produced. By the time that adult life is reached, these muscle fibres have increased seven-fold and yet further enlargement can occur in response to chronic overload such as is found in lung diseases or systemic hypertension [244]. Hypertrophy of cardiac muscle cells during postnatal development may involve nuclear division in which cells become binucleate while later in life the nuclei fuse to produce polyploid cells [99]. Increasing polyploidy is also a feature of liver tissue and may be related to the hypertrophy of individual cells [6].

Many types of hyperplasia increase the functional capacity of the organ and have a beneficial value. Lymphoid tissue undergoes hyperplasia in chronic infection and following antigenic stimulation, one example of this being the enlargement of the spleen in malaria [244]. Oxygen deficiency due to prolonged exposure to high altitude is compensated for by increase of up to 50% in the red cell count [45]. In addition to these conditions of 'physiological hyperplasia' there are many states of 'pathological hyperplasia' in which the increase in cell number has no apparent adaptive advantage. One such condition is psoriasis, which is a skin disorder involving a massive increase in the rate of epidermal cell division [242] and an expansion of the germinative zone to three layers of proliferating basal cells [241].

It should be noted that hyperplastic growth is self-limiting in that the balance between cell production and loss is still maintained though at a different rate of turnover.

### G. Oedema

Mechanical or chemical injury is often followed by a very rapid increase in volume of the surrounding tissue. Immediately after injury, a protein-containing exudate accumulates due to the escape of fluid from damaged capillaries. These proteins, and their breakdown products, exert an increased osmotic pressure which further encourages fluid retention. This process is termed 'oedema' and does not appear to involve any significant change in cell size or number [85].

### H. Neoplasia

In neoplastic conditions (cancers or tumours) the stabilisation typical of normal cell turnover or of hyperplasia breaks down and the result is a progressively growing mass of tissue. There is some evidence that neoplasia in some tissues may originate in areas which have already undergone some form of hyperplastic change [226]. Neoplastic cells almost invariably show some loss or change in normal physiological function in addition to disordered growth. One example of this is the 'oat cell' tumour of the lung in which the loss of differentiated function is so great that it is uncertain whether it derives from connective tissue or from the epithelium [244].

Although the cells of many tumours proliferate more rapidly than the cells of the normal tissue from which the tumour originated, the duration of the cell cycle in many normal tissues can be shorter than that of the most rapidly growing tumours. It appears that the size of the growth fraction and the rate of cell loss are more important than the rate of cell division in determining the growth of normal and neoplastic tissues [142]. Cultures of neoplastic cells can yield apparently normal cells which are incapable of forming tumours. This suggests that the normal cell genome contains all the information required for the expression of the neoplastic phenotype and that neoplasia is due to a faulty developmental process [172], possibly involving some increase in the stability of mRNA whose products regulate the rate of cell turnover [173].

A distinction is often made between benign tumours which are slow-growing and non-fatal and malignant tumours which have a rapid rate of growth and which usually are fatal unless treated [7]. While this distinction is helpful from a clinical point of view, it is uncertain whether there is any true biological difference between the two categories or whether they simply represent different extremes of the same phenomenon of disordered growth.

### I. Cell Loss

In addition to the cell loss observed in renewing tissues and referred to above, tissue degeneration due to massive cell loss is a common phenomenon during embryogenesis and developmental growth [65] and examples may be found in most organ systems of the body [92]. It occurs during the pupal stage of development in insects where larval tissues degenerate and are replaced by the proliferation of cells in the imaginal buds [250] and it is also observed in the development of Amphibia where the larval kidney or pronephros completely disappears, to be replaced by the mesonephros or adult functional kidney [166]. The cytological appearance of degenerating cells in tissue renewal and in developmental cell loss is very similar and it has been proposed that a common mechanism, termed 'programmed cell death' or 'apoptosis', is involved [123].

### III. The Relative Importance of Cell Programming and of Regulation

The question arises as to whether the various growth phenomena listed above can be explained on the basis of a fixed intracellular programme or clock which determines the rate at which cells divide and the number of divisions a cell passes through before death or loss or whether these parameters can be modified by regulatory systems involving interaction between cells, between tissues or organs or between the organism and its environment. Observations designed to test the relative importance of programming and regulation were carried out on the intestinal epithelium. In this tissue there is a high rate of proliferation in the stem cells of the crypts but the cell population remains constant due to migration of cells and loss at the tips of the villi [257]. One way of explaining this equilibrium [169] is that it might be due to a cellular programme which determined that, after mitotic division of the stem cell, one daughter remained a stem cell while the other became a 'maturing' cell which was eventually lost. However, it can be shown that in addition to this 'unequal' division, some pairs of daughter cells both remain stem cells while others both progress to maturing cells. This evidence strongly indicates the existence of regulatory processes [47].

The concept of programming cannot, however, be dismissed. Cells from rapidly growing foetal tissues typically retain a high rate of division when maintained in tissue culture as compared with homologous cells from mature tissues [109]. For example, the rate of mitosis in alveolar cells of mouse lung maintained in organ culture for 3 weeks remained higher than in equivalent cells from adult lung kept under identical conditions [216]. This evidence strongly supports the idea of cellular programmes.

Another aspect of intracellular information which, in many situations, seems to be an important determinant in cell division is the cytoplasmic mass [260] or the cytoplasm/DNA ratio. The decrease in cell division rate after the first few cleavage divisions during embryonic development appears to be programmed and it has been suggested [190] that the change is caused by the cytoplasmic/nuclear ratio falling below a critical level. Growth of many insect tissues is due to an increase in cell size rather than to cell division, an observation which seems inconsistent with the concept of cell division being triggered by growth to a critical size. However, in such insect cells there is a high degree of polyploidisation which may correspond to over 10 duplicative steps and the effect of this is to keep the cytoplasm/DNA ratio relatively constant [61]. In some cell systems, some critical amount of $G_1$ growth is necessary for the induction of the S phase and theoretical studies suggest that the timing of mitosis may be dependent on cytoplasmic mass [81]. Work on yeast cells suggests that the attainment of a critical cell size is, in itself, unlikely to be the direct stimulus to division but that the important factor is some cellular property which changes with size such as the accumulation of some specific cell component [51]. A similar process has been proposed to account for the lengthening of the cell cycle in early embryonic tissues, namely that preformed cytoplasmic deoxyribonucleotides, which act as a reserve for rapid DNA synthesis, constitute a critical component whose exhaustion determines a reduction in the rate of cell division [133].

It thus appears that cell division rates are, at least in part, determined by intracellular information or programmes which are modified by the process of cellular ageing. The concept of cell ageing has been developed mainly on the basis of studies on cultured diploid cells and there is evidence to suggest that some mammalian cells have a finite life span and that, having passed through a specific number of consecutive mitotic divisions, they are programmed to die [225]. Whether this is generally true for other species and cell types is uncertain, since other types of cultured diploid cell, notably those of Amphibian origin [178], appear to have a much higher limit. However, while the idea of a finite number of mitotic divisions may be open to question, there can be no doubt that cells do age or senesce and that this is due to the progressive effects of cell division. In the development of Amphibian embryos each stage of differentiation is determined by the cells achieving a specific number of divisions [35] and a similar phenomenon has been described in the development of appendages in the chick [256]. It has been suggested [114] that cells count the number of divisions by successive methylation of adenine in DNA and that the process may be reversible, either by the action of a demethylating enzyme or by division in the absence of the methylating enzyme. Other workers have proposed that cell ageing may be due to the accumulation of peroxidized lipids in the cell membranes [205] and cell fusion experiments also favour the idea that at least part of the cell ageing mechanism may reside in the cytoplasm [162].

The control of apoptosis or programmed cell death is believed to involve a different mechanism from that which regulates the cellular ageing phenomena referred to above [144]. Cells from the posterior necrotic zone of the chick wing bud, isolated as early as their location can be determined, will undergo apoptosis even when cultivated in vitro, which suggests that the programming must occur very early in development [80] though, in its earliest stages, the effects may be reversible [144]. The signal which initiates programming may induce the synthesis of specific types of mRNA but the activation of the programme, leading to cell death, may involve control at the post-transcription level [61]. Experiments on thyroxine-induced tail regression in Amphibia suggest that this hormone acts by inducing the initial activation of enzyme cascades in which one enzyme stimulates the production of larger quantities of the succeeding enzyme [22].

From observations referred to above, it seems that changes in cell division rate during the earliest stages of development are due to cell programming and it is therefore relevant to ask at what stages of development do supracellular regulatory mechanisms first come into play. Unilateral removal of the Amphibian pronephric kidney at an early stage of development leads to a compensatory increase in the rate of cell division in the remaining pronephros [54] while damage to the kidney of mature Amphibia induces an increase in the rate of cell division in the pronephros of implanted embryonic grafts [55]. Growth inhibitory factors prepared from adult tissues can affect the rate of cell division in embryonic liver [36, 49, 138] and intestinal epithelium [37]. It would, therefore, appear that even the rapidly dividing cells of embryonic tissues are subject to some form of supracellular regulation.

It seems, therefore, that the rate of cell division in differentiated tissues is determined by two factors: a cellular programme which is modulated by supracellular regulatory mechanisms. The main part of this review is devoted to a discussion of the possible nature of the regulatory processes.

## IV. Common Factors in the Control of Cell Division

The immediate stimulus to cell division lies in the activation of specific biosynthetic pathways which are probably under genetic control. Theories for the control of cell division must therefore explain how intracellular and supracellular regulatory factors interact with the genetic apparatus of the cell. Some authors [39] envisage the supracellular regulatory factors as acting directly upon operons which are linked to the genes whose activity is essential for cell division. It now appears more probable that cell division is controlled by the supracellular factors binding to specific cell surface receptors from where the signal is relayed to the genetic apparatus through a different mechanism [28]. This 'second messenger' concept may be defined as a chemical signal generated by the interaction of an external stimulus with a cell surface receptor in order to initiate the appropriate cellular response [180]. An example of this phenomenon is in the action of many hormones which act at the cell surface, the message being carried intracellularly by a different substance. For a variety of cellular control processes there is now much support for the view that the cyclic adenosine monophosphate/cyclic guanosine monophosphate (cAMP/cGMP) system constitutes the second messenger. Earlier work favoured the view that cAMP was the active agent. It was noted that in some cell systems an increase in the rate of DNA synthesis was accompanied by a decrease in intracellular cAMP [168], and it was also shown that DNA synthesis could be inhibited by the addition of extracellular cAMP [90]. In other cell systems it was subsequently found that DNA synthesis and cell division could be initiated by an increase in the level of cGMP [93, 259]. It now seems probable that cell division is regulated by the joint action of cAMP and cGMP and that their relative importance may differ according to the particular cell system [139, 180]. For example, in wound healing cGMP appears to transmit the proliferation signal while cAMP promotes differentiation [165]. However, it is by no means certain that the cAMP/cGMP directly controls gene activation since there is much evidence that intracellular calcium ($Ca^{2+}$) is also involved in the regulation of cell division [32, 180]. Increased levels of cAMP cause the immediate release of $Ca^{2+}$ ions from mitochondria, where most of the intracellular calcium is stored, and it has been proposed that this might regulate cell division [31]. Mitochondria contain specific binding proteins which act as receptors for cAMP [11] and also their inner membranes contain a specific $Ca^{2+}$ carrier [46] through the interaction of which mitochondrial gene transcription and many other macromolecular biosynthetic pathways, including those required for cell division [180], might be regulated.

The interaction between cAMP and $Ca^{2+}$ appears to be mediated by the modulator protein calmodulin [58]. Binding of the cAMP to an inactive holoenzyme present in the cytoplasm causes its dissociation into an active protein kinase which promotes the phosphorylation of certain proteins which then modulate various physiological responses in the cell. One of these modulator proteins is calmodulin, which binds with $Ca^{2+}$ to form an active complex. Calmodulin also seems to mediate in the activation of synaptosomal kinases found in nerve endings [118] and it may thus be involved in the release of neurotransmitter substances [73] which, as will be shown later, may participate in the regulation of cell division.

It is now widely believed that intracellular messengers do not act directly upon the genes controlling cell division but that their effect is mediated by genes which govern other aspects of cell function. There is much evidence to suggest that, within a single cell, mitotic activity and the expression of other differentiated cell functions tend to be mutually exclusive [39]. Theoretical studies suggest that functional and mitotic operons are connected in such a way that activity of one group produces a metabolic repressor of the other group [239]. This concept is relevant to the induction of neoplasia, which may be considered to be an aberration of cell division involving an altered pattern of gene expression in stem cells [172]. Neoplasia may therefore involve the blocking of functional operons with the consequent derepression of mitotic operons [240].

The nature of the genes involved in cell division must now be considered. In mammalian cells only a small proportion, probably less than 5% of the total chromosomal DNA, is concerned with the processes of cell growth and division and, from work on yeast cells, there is no evidence to suggest that these genes are closely clustered on the genetic map. In yeast cells, division appears to be regulated by two parallel but separate genetically controlled enzymatic sequences, one concerned with DNA synthesis and nuclear division and the other involved with cytokinesis [177]. The enzymes involved in nuclear replication belong to two main categories: those such as chromosomal NAD-pyrophosphorylase and poly (ADP-ribose) polymerase which govern the energy requirements for DNA synthesis [203] and those which are required for the process of DNA replication itself. The latter group includes DNA ligase stabilizing factors, DNA polymerase [130], DNA ligase, helix-destabilizing factors and deoxyribonuclease which promotes unwinding of the double helix [255]. The enzyme sequence involved in cytokinesis is responsible for the synthesis of sulphydryl-rich proteins which constitute the mitotic apparatus and determine its contractile abilities [183, 199].

The nuclear division and the cytokinetic pathways are normally closely coordinated so that DNA synthesis normally appears to lead inevitably to mitosis. However, when many tissues such as lung [208], liver and kidney [212] are maintained in organ culture this coordination breaks down and there is an excess of DNA synthesis over cell division leading to the production of large numbers of tetraploid cells. This may have a bearing on the phenomenon of cell hypertrophy, in many forms of which there is an increase in the degree of polyploidy.

In eukaryotic cells the biosynthetic pathways leading to cell division are remarkably similar and it is reasonable to suppose that there is some intracellular and genetic control mechanism which is common to different species, cell types and growth phenomena. Many authors have also proposed unified theories for the control of cell division at the supracellular level which provide a common basis for explaining such phenomena as developmental growth, cell turnover, various forms of compensatory growth [39, 248] and neoplasia [117]. However, the notion that a supracellular signal may be translated by cell surface receptors [28] into information carried by a second (intracellular) messenger [180] allows the possibility that supracellular regulatory factors may be of many types, varying according to the nature of the growth phenomenon. In this review we shall be particularly concerned with the control of growth following tissue damage or loss.

## V. Tissue Response to Trauma

Traumatisation of a tissue by physical or chemical agents or by infection produces a rapid response within a few seconds of the application of the stimulus. Blood vessels in the immediate vicinity dilate leading, in the case of surface tissue such as skin or conjunctiva, to reddening or erythema. The permeability of the blood vessels also increases, permitting extravasation of fluids and proteins which leads to local swelling or oedema and the area also becomes a focus for migratory white blood cells, particularly neutrophil polymorphonucleocytes which may accumulate to the extent that they form pus [244]. Such a response may be observed in a wide variety of tissues and the different elements of the response are referred to collectively as 'acute inflammation'.

Mediation of the acute inflammatory reaction is a complex process involving chemical agents and neural mechanisms. The immediate effect of tissue damage appears to be two-fold [254]. One sequence involves the liberation of proteolytic enzymes from damaged cells which results in the formation of modified proteins with new antigenic activity. This activates the complement system and also causes antigenic stimulation of lymphocytes. The other sequence involves activation of the blood coagulation system leading to the formation of kinins [161], the agglutination of platelets with the consequent release of histamine and the formation of fibrin. Activation of the complement system initiates a complex set of events [188] which include liberation of the vasoactive amines histamine and serotonin [224] from mast cells [251], the production of complement fragments which act as chemotactic agents controlling the migration of white blood cells and probably the release of prostaglandins and kinins [254].

Damage or irritation also generates an impulse in local sensory nerve endings which not only travels back in the normal direction towards the central nervous system but also enters nearby branches of the same nerve, travelling in the reverse direction (antidromically) to local blood vessels, causing them to dilate [156]. This phenomenon is termed the 'axon reflex'. Stimulation of the nerve endings by the damaged tissue is probably mediated by histamine while the dilatation of blood vessels by the activated nerve probably involves bradykinin [86]. A further function of bradykinin may be to cause increased permeability of the blood vessels [254].

There is still no universal agreement as to the exact way in which acute inflammation is mediated and many of the processes described above are open to question. Some authors [115] believe that vasodilatation is mediated by E-type prostaglandins and that the function of bradykinin is restricted to mediation of increased vascular permeability [252]. The importance of the axon reflex has been questioned and it has been claimed that its function may be to cause degranulation of mast cells and the liberation of histamine, leading to increased vascular permeability and oedema [124]. If, however, we neglect the controversy as to their exact function, there does seem to be general agreement upon the importance of certain chemical mediators, particularly the kinins (e.g. bradykinin), the prostaglandins, histamine and possibly serotonin (5-hydroxytryptamine). It also seems to be very widely accepted that the nervous system is involved [156] and that tissue damage results in the appearance of new antigens which stimulate the production of new antibodies.

Tissue damage, provided it is of sufficient extent, can give rise not only to a local reaction but also to changes in homologous tissue situated at some distance from the site of injury. Impairment of lung alveolar tissue may cause vasodilatation and an increase in the rate of blood flow in other parts of the lung lobe or even in other lobes [213]. Partial hepatectomy has also been found to increase blood flow in the remaining parts of the liver [24] while after unilateral nephrectomy there is a rapid transitory increase in blood flow in the remaining kidney [38]. This phenomenon may be a response to the increased physiological load placed upon the remaining tissue since, even in the absence of tissue damage, when a resting organ becomes active its blood flow is rapidly and greatly increased, probably due to the vasoactive effect of plasma kinin [86]. Such haemodynamic stresses may cause aggregation of platelets and deposition on the walls of the capillary endothelium [20, 76] leading to thrombin-induced release of vasoactive amines [94]. A rise in the rate of blood flow also causes back pressure from the venous system which increases the rate at which proteins escape through the capillary wall into the interstitial spaces [89]. Neural mechanisms probably also play an important part in this phenomenon since, as will be discussed in greater detail later, vasoconstriction and vasodiltation are regulated by the autonomic nervous system. It can therefore be seen that these responses to tissue injury may be mediated by chemical and neural mechanisms very similar to those which control the local inflammatory reaction.

## VI. Cell Division and Tissue Trauma

In situations where tissue damage or loss is followed by compensatory growth or regeneration (see above for definitions) there is a rapid increase in the rate of cell division. Within about 6 h of the original stimulus there is an increase in the rate of RNA synthesis and soon after an increase in protein synthesis [24]. This process probably represents the transcription of mRNA and the synthesis of enzymes required for entry into cell division. By about 12 h the rate of DNA synthesis has begun to increase, due to cells entering the S phase of the cell cycle and within a further 2–6 h the rate of entry into mitosis (M) is also on the increase. The maximum rate of DNA synthesis is achieved by 24–48 h with the maximum rate of entry into M a few hours later. This lag is because, once DNA synthesis is completed, the cell has to pass through the $G_2$ phase before entering M. The maximum rates of DNA synthesis and mitosis last for only a few hours after which there is a sharp decline so that by about the 5th day these rates have returned almost to their normal resting level. Restoration of the lost tissue mass usually takes between 2 and 4 weeks and is achieved by the daughter cells, produced by mitosis, growing up to the size of the original parent cell. The times given above are approximate since they vary according to the cell type and often also to the age of the organism. However, the pattern of response is remarkably constant for a wide variety of organs [39, 98] including the epidermis [39], kidney [38], lung [84] and liver [24]. Some ideas as to how this response might be controlled will now be considered.

## VII. Theories on the Control of Cell Division following Tissue Trauma

Various possible mechanisms for the control of cell division at the intracellular and genetic level have already been discussed. We will now review some of the principal theories concerning regulation at the supracellular level. Many of these are general theories in the sense that they seek to explain a range of growth phenomena in addition to that which follows tissue damage.

### A. Wound Hormones

It has been established that traumatized cells of higher plants liberate a substance which stimulates cell division in nearby cells, thus contributing to the process of regeneration. The pattern of response to local tissue damage in plant cells in many ways closely resembles that in animal cells [69] and it has been proposed that division in animal cells may also be regulated by stimulatory wound hormones [1, 229, 230]. Many growth-promoting substances liberated from traumatized animal cells are nucleic acid precursors and may be regarded as nutritive factors rather than true growth regulators [69]. However, evidence from a variety of cell types, including orbital gland, liver, granulocytes and erythrocytes, suggests that the effect of cell breakdown products may be tissue-specific which indicates that they may be part of a natural regulatory mechanism [232]. Experiments on damaged kidneys [10] and resected bowel [253] suggest that the release of stimulatory substances from traumatized cells may be important in these organs also. We shall return to the question of wound hormones in the discussion of the effects of inflammation on cell division.

### B. Growth Regulation through Negative Feedback

In contrast to the above theory, which proposes that cell division is regulated by stimulatory substances, it has also been proposed that division is controlled by a process of negative feedback involving autoregulative mitotic inhibitors produced by the tissues themselves. This theory proposes that the rate of division is normally held in constraint by the inhibitor but that a deficit of cells, from whatever cause, leads to a reduction in the concentration of the inhibitor so that the rate of division can increase. A mathematical model of this hypothesis of growth regulation has been described in which it was suggested that the exact mode of action of the inhibitor or 'antitemplate' was to neutralize the stimulatory effect of an intracellular messenger or 'template' [248]. An important part of this model, and of similar hypotheses, is the assumption that, in order to explain the apparent tissue specificity of growth regulation, each tissue must have its own chemically specific inhibitor.

Support for this hypothesis came from experiments in which intraperitoneal injection of homogenates from liver or kidney produced organ-specific inhibition of cell division in the corresponding organs [197]. Experiments on the epidermis, designed to test whether the increased mitotic rate after tissue damage was caused by the liberation of stimulatory wound hormones, suggested that the effect was due to a deficit of a mitotic inhibitor or 'chalone' [39]. The validity of this conclusion will be discussed later. Further support for the existence of an epidermal chalone came from experiments in which tissue-specific inhibition of mitosis was obtained by the addition of tissue homogenates to cultures of epidermis [41]. Existence of chalones or tissue-specific growth inhibitors has since been claimed in a variety of tissues [39], including melanocytes [159], granulocytes [194], lung alveoli [215], kidney [38, 56]. fibroblasts [116] and liver [138]. Cell proliferation in the intestinal epithelium is controlled by a feedback effect [185] mediated by a putative chalone and two endogenous inhibitory hormones, secretin and pancreozymin [25].

There is evidence to suggest that the rate of division may be regulated at more than one point in the cell cycle and that there might be two types of chalone, one controlling the movement from $G_1$ into S and the other from $G_2$ into M. This dual system has been described in a number of tissues, including intestinal epithelium [37] and epidermis [234]. Differences in the activity or nature of chalones between slowly and rapidly growing phases of the same tissue have also been described. The chalone-like inhibitory effect is reduced in extracts from rapidly proliferating cirrhotic or regenerating liver [145] while, again in the liver, there may be an inverse correlation between the inhibitory action of the tissue homogenate and the mitotic activity in the donor organ according to the time of day at which the sample is taken [49]. Experiments on Amphibian tissues suggest that extracts of adult organs inhibit cell division in the corresponding larval organs [36, 56] but other evidence suggests that adult organ extracts are less efficient at inhibiting the activity of the more rapidly proliferating immature organs. The $G_1$ chalone extracted from adult epidermis does not inhibit DNA synthesis in the neonatal epidermis [149] and it has been proposed that immature tissues contain specific growth regulators different from those present in adult tissues [263]. Although they are rapidly proliferating, the cells of neoplastic tissues do seem responsive to inhibitory extracts from normal tissues. This phenomenon has recently found a practical clinical application in inducing regression of granulocytic leukaemia [195].

The existence of chalones is, however, by no means generally accepted. Destruction of normal tissue by neoplastic invasion does not initiate cell division, which suggests that chalones cannot control proliferation by a simple process of negative feedback [9]. A similar conclusion follows from the consideration that, after the initial burst of cell division following tissue damage, the mitotic rate returns to normal well before tissue mass is restored [24]. Despite considerable research effort, the chemical identification of chalones has proved to be elusive, progress in most directions being limited to the identification of active fractions from tissue homogenates and estimations of the molecular weight using gel or membrane molecular filtration [116]. It has therefore been suggested that the biological effects of tissue extracts may be due not to the presence of putative chalones but to the action of spermidine, a low molecular weight polyamine [75].

## C. Growth Regulation by Physiological Load

Damage or loss of tissue throws an increased physiological burden on the remaining homologous tissue and it has been proposed that this constitutes a direct stimulus to growth, the effect possibly being mediated by the accumulation of some chemical factor which is normally metabolized or excreted by the tissue concerned [98]. This theory has much to recommend it, especially in view of the fact that renewed growth can be regarded as an evolutionary adaptation designed to restore normal physiological functioning [98]. In relation to this theory, much interest has centred on the kidney, where compensatory growth, following unilateral nephrectomy, has been thought to be due to the increased excretory function imposed upon the remaining tissue. However, doubling the excretory load on one kidney by diverting the ureter of the contralateral kidney does not stimulate growth and this observation has been used as an argument against the theory. The reasoning may, however, be invalid [121]. The main function of the kidney is regulatory rather than excretory since a large proportion of the metabolites removed from the blood by glomerular filtration are reabsorbed elsewhere in the kidney. Ureteric diversion increases the total load by only 1%, which is considered insufficient to induce growth.

Some evidence in support of the theory comes from experiments on the liver. Following cross-linkage of the blood supply in two animals (parabiosis), partial hepatectomy in one partner stimulates liver growth in the other, which strongly supports the idea of control mediated by a circulating humoral or blood-borne factor [198]. This factor appears to reach the liver via the portal supply and it is thought that it does not originate in the liver itself. [83]. It is not therefore an autoregulative factor akin to wound hormones or chalones but could possibly be a product of digestion normally metabolized by the liver. Experiments on intact animals demonstrate that fatty acids, which may originate through digestion of lipids, can induce DNA synthesis in the absence of tissue damage and these results are supported by the fact that DNA synthesis can be stimulated by glucagon which increases that supply of fatty acids to the liver [206]. Short chain fatty acids derived from fibre digestion stimulate epithelial cell proliferation in the rumen, probably through the mediation of gastrin and insulin [200], though in the skin, increased cell proliferation may be related to a deficiency of fatty acids [154].

*D. Possible Function of the Lymphoid System in Growth Regulation*

There are several aspects of the lympoid system which recommend it for consideration as part of the growth regulatory mechanism. It is capable of detecting the level of circulating antigens and thus monitoring the amount of tissue synthesizing such antigens and it is capable of responding by the synthesis of highly specific macromolecules. An ingenious theory has been proposed [43] in which tissues are thought to release highly specific tissue coding factors (TCFs) in response to which the lymphoid system produces specific mitotic control proteins (MCPs) which inhibit cell division in the corresponding tissue. Loss or damage to the tissue would cause a reduction in the level of the circulating TCF, a reduction in the amount of MCP produced and hence a release from mitotic inhibition.

Most of the evidence supporting the idea of lymphoid involvement in growth control comes from the liver, but in this case it seems that the action of lympoid cells is stimulatory rather than inhibitory. Lymphoid cells from partially hepatectomized mice stimulated cell division in the liver of unoperated mice [14], while the rate of production of antibody-forming cells is higher in strains of mice which have a higher rate of liver regeneration [16]. Partial hepatectomy is accompanied by increased activity in thymus-dependent parts of the spleen and in the bone marrow [127]. Liver regeneration following partial hepatectomy is slower in irradiated mice, possibly because of impaired lymphoid production [70].

Amphibian kidney extracts which had a chalone-like inhibitory activity were injected into rabbits to produce an antiserum which, when added to cultures of larval kidney (pronephros), stimulated cell division [57]. This experimental situation may be likened to Grave's disease in which hyperplasia of the epithelial element of the thyroid gland is associated with two species of autoantibodies, the long-acting thyroid stimulator (LATS) and the human thyroid-stimulating immunoglobulin (HTSI) [135]. In the corneal epithelium, the lymphoid system appears to promote entry of cells into mitosis during the diurnal cycle of mitotic activity since splenectomy causes a reduction in the daily peak of cell division [148]. While there is considerable evidence to support the view that the lymphoid system participates in growth regulation, the data do not confirm the idea of TCFs and MCPs described above [43].

### E. Control of Growth by the Nervous System

The central nervous system is a highly efficient monitoring device capable of receiving information from all parts of the body via sensory receptors and efferent nerves and of sending back specific messages via the afferent nerves. It would seem likely that damage or functional impairment of tissue should be registered by the central nervous system and there is evidence that neural mechanisms mediate in many regenerative phenomena.

The size of the normal salivary gland appears to be regulated by action of the sympathetic nervous system [98, 99, 106]. Section of the sympathetic innervation of this organ causes atrophy and prevents compensatory growth following tissue damage. Chronic sympathetic nervous stimulation of the salivary gland by repeated amputation of the lower incisor teeth causes growth of the organ. Adrenergic-blocking agents such as dibenamine, which interfere with neurotransmission (see below), prevent compensatory growth while sympathomimetic agents such as isoproterenol stimulate growth.

Following amputation, the limbs of many Amphibian species regenerate completely. Regeneration does not occur if the limb is denervated and it appears that nerves release a neurotrophic agent which stimulates cell division. Culturing in association with dorsal root ganglion tissues stimulates the growth and differentiation of regeneration blastema tissue [160], while diversion of nerves to an amputated limb induces growth of a regeneration blastema even in species which normally do not regenerate lost limbs [221]. Implantation of nervous tissue stimulates growth and differentiation in digits of amputated limbs even in mammalian species where extensive limb regeneration is normally unknown [158]. One aspect of the nerve-dependent changes in newt limb regeneration is that the neural stimulus may promote protein synthesis by increasing the number of polysomes engaged in mRNA translation [19].

Attemps at isolating the agent responsible for limb regeneration suggest that, in addition to the normal neurotransmitters such as noradrenaline and acetylcholine, other trophic substances may be involved [221]. Noradrenaline stimulates β-receptors in blastema cells and thereby raises the intracellular level of cAMP [181]. However, this short-term excitatory effect does not appear to be the cause of the increased mitotic activity in blastema cells. Growth of the blastema seems to be dependent upon long-term neurotrophic effects mediated by peptide messengers which regulate the production of a number of enzymes including acetylcholinesterase [182].

The β-adrenergic system, through its action on cAMP, may also mediate in the regulation of epithelial cell division and it has therefore been proposed that the growth of this tissue is under neural control [242]. Regeneration of the Amphibian lens from the iris may be mediated by neural mechanisms. When implanted into denervated muscle, iris tissue forms fewer lens vesicles than when grafted into normal muscle [184]. Regeneration of skeletal muscle may likewise be dependent on neurotrophic agents. Normal muscle grafted into animals with muscular dystrophy tends not to regenerate, while dystrophic muscle implanted into normal animals has an improved regenerative capacity [111]. These results may be due to defects in the nerve supply of dystrophic muscle, since in dystrophic animals the axons supplying the affected skeletal muscle are reduced in number and are abnormally myelinated [33].

A number of human growth disorders may involve some abnormal function of the autonomic nervous system. Psoriasis, and a number of other forms of hyperplastic skin disorders which involve increased cell multiplication or disturbed growth homeostasis [242], may be caused or exacerbated by psychogenic factors such as anxiety and other chronic emotional distress [228]. A possibly related effect is the development of inflammatory lesions which may appear under the influence of hypnotic suggestion [52]. Psychogenic distress appears to be channeled into organic dysfunction through the mediation of the hypothalamus and autonomic nervous system [113] and this may lead to disorders of the endocrine system [247] and of cell proliferation [119].

*F. Biological Factors Affecting the Growth of Cells in Culture*

A number of factors have been identified which stimulate the proliferation of cells in culture [96]. They include the somatomedins which comprise a family of peptides synthesized under the influence of the pituitary growth hormone and which promote division in a number of different cell types. Also important are certain specific growth-promoting serum proteins and growth factors from various tissues, including nerve (NGF), epidermal (EGF), ovarian and fibroblast growth factors [96]. All these agents have been isolated in chemically pure form and all have some degree of tissue specificity in their action. However, some degree of cross-reaction can occur, for example myoblasts can be stimulated to divide by fibroblast growth factor (FGF) while chondrocytes respond to both EGF and FGF [95]. It appears that FGF is specific to the extent that its mitogenic action is confined to mesoderm-derived target cells [97]. It is possible that these growth factors may participate in the natural growth control mechanism either by stimulating cell division or, as appears to be the case with EGF, by promoting the expression of mature differentiated characteristics [131] and protecting against programmed cell death [144].

Human diploid fibroblasts cultured at a high density produce two substances which are both distinct from the fibroblast growth factor and which promote the growth of cells maintained at low density [155]. These agents appear to act by binding to specific mitotic control receptors which are believed to exist on the cell surface [28].

*G. Action of Specific Hormones on Cell Division and Growth*

In endocrine organs and endocrine target organs such as secondary sexual organs, growth is promoted by hormones which have a specific action on the tissues concerned [134]. There is some evidence that specific hormones may also affect tissues which are not part of recognised endocrine target organs. The growth and cell division of the salivary gland, for example, is affected by a number of hormones, some of which have a stimulatory effect (pituitary growth hormone, thyroxine and insulin) while others, such as the parathyroid hormone, have an inhibitory effect. Growth of the kidney is likewise promoted by a number of hormones which include thyroxine and testosterone, while other hormones such as oestrogen act as inhibitors [99]. Although parathyroid hormone appears to inhibit salivary gland growth [21] it may be essential for the proliferation of liver cells after partial hepatectomy [187]. Certain hormones, in particular prolactin, may also be important in promoting cell proliferation in the gastrointestinal tract [204].

A complicated state of hormonal dependence exists in the salivary gland, which is stimulated to grow by the excision of the lower incisor teeth, but only if the testis remains intact [15]. This suggests that testicular hormones, in addition to those described above, are important in mediating the growth of this organ.

In some species, notably mice [48] and frogs [211], the rates of cell division in a variety or organs within an individual animal tend to correlate: if a high mitotic rate occurs in one organ it is usually observed in other organs. This suggests some form of hormonally mediated organismal control. Adrenaline has an inhibitory action on cell division in many tissues and it has been proposed that this hormone acts as a controlling agent at the organismal level, possibly by potentiating the action of putative organ-specific chalones [39]. This idea is open to doubt since, when tested on a wide variety of organs in culture, adrenaline appeared to have no consistent effect upon cell division [212].

Certain critical stages in the differentiation and proliferation of haematopoietic cells appear to be controlled by specific hormones. Erythropoietin stimulates the production of erythropoietic stem cells [13] while specific colony-stimulating factors have a similar effect during granulopoiesis [44]. These substances can be isolated from the urine of experimental animals and are probably part of the natural regulatory mechanism.

### H. Growth Control by Tissue Interaction

Most organs contain a number of tissue or cell types. It is difficult to imagine how the growth and accurate determination of the relative numbers of different cell types could be regulated solely by autoregulative processes and it seems more plausible that controls involving interactions between different cell types also exist. One possible model for such a control is that one tissue produces a stimulatory factor specific for an adjacent tissue [164]. If, during the growth process, an imbalance developed due to excess cell production in one tissue there would be an increase in the quantity of the stimulatory factor which would then promote more rapid growth of the second tissue. Growth-promoting action between different tissue types is observed during organogenesis [104] and this may be the basis for the phenomenon of induction, which is important in the development of many embryonic organs including the liver, kidney [17] and lung [222]. It is possible that induction, which has been studied mainly in developing systems, may also be important in maintaining the dynamic equilibrium between different cell types in mature organs.

In some respects tumours resemble normal organs in that they contain a variety of cell types including the true neoplastic cells, normal connective tissue cells and the various elements which make up the blood vessels. Growth of the tumour involves some degree of coordination between normal and neoplastic cells, since, if the neoplastic cells multiply in excess of the blood vascular network, the tumour tissue may become necrotic due to ischaemia. It appears that adequate growth of the vascular supply is regulated by tumour angiogenesis factors, produced by neoplastic cells, which stimulate the proliferation of endothelial cells over a distance of up to 5 mm [87].

*I. Some Common Assumptions Shared by Hypotheses of Growth Regulation*

With the exception of the postulated neural control mechanisms, all the hypotheses discussed above share certain common ideas. Cell division is envisaged as being controlled by some diffusible intercellular chemical signal. Since, in many cases, homologous tissue at some distance from the site of tissue damage can respond by an increased rate of cell proliferation, it is further assumed that such chemical signals must be distributed systemically as 'humoral' or blood borne factors. Finally, because of their systemic distribution, it is necessary to postulate that the factors are tissue-specific to avoid the possibility of damage in any given tissue stimulating cell division in other tissue types. Despite their many dissimilarities, most hypotheses therefore assume the existence of tissue-specific, systemically distributed chemical regulators [213]. Experiments will now be described which cast some doubt on this idea.

*VIII. Regulation of Growth in Lung Alveolar Tissue*

Following partial removal of the lung, for example resection of the whole left lobe in rodents, there is a typical compensatory response involving a massive increase in the rate of cell division [84, 189]. This could be explained by any one of a number of hypotheses. For example, it could be due to the liberation of wound hormones from the damaged residual tissue of the resected lobe, to the increased physiological respiratory load placed on the remaining alveolar tissue or to a reduced systemic 'lung chalone' level due to reduced tissue mass. It has been shown that tissue-specific inhibition of cell division in lung organ cultures can be obtained by addition of tissue extracts which could be interpreted as supporting the existence of a lung chalone [215]. Stimulation of lung alveolar cell proliferation can also be obtained by simple collapse of the contralateral lung following injection of air into the thoracic cavity [210]. Involvement of putative wound hormones is excluded, since no trauma to lung tissue occurs as the result of experimental treatment and since there is no loss of tissue mass, the results cannot be explained on the basis of a reduction in the level of a putative lung chalone. It would seem that the results are consistent with the idea of mitotic stimulation through the increased respiratory load placed on the remaining functional alveolar tissue. However, packing the thoracic cavity with a sterile implant prevents growth [62] and mitotic stimulation in the lung following removal or collapse [210] of the contralateral lobe. This lack of response, occurring in the presence of an increased respiratory load, indicates that physiological stimulus cannot be the prime regulatory factor.

The above experiments cast doubt on some of the principal hypotheses concerning the chemical nature and mode of action of the proposed regulatory factors and further experiments also call into question the concept of systemic distribution. Small localized wounds made by cauterizing a small area on the lung surface stimulate a reparative response which involves increased cell division in the nearby intact alveolar tissue [214]. The amount of tissue destroyed in this operation is less than 0.25% of the total lung tissue and a change in the systemic level of a regulatory substance, due to reduction of tissue mass, can hardly explain such a marked growth response, particularly since this is confined to the immediate vicinity of the damaged tissue.

In the case of the experiments on unilateral pneumonectomy or lung collapse, creation of a vacant thoracic space leads to a marked mediastinal shift in which the whole heart-lung complex becomes displaced to one side [62, 84] leading to hyperinflation and a decrease in the physiological dead space in the remaining functional lung lobes [64]. A growth response occurs in all cases where this movement is allowed to take place and is absent in situations where the shift is prevented by mechanical obstruction with a thoracic implant. Removal or collapse of pulmonary tissue is followed by diversion of blood flow from the functionless areas to other parts of the lung [192] and it was proposed that the increased blood flow would cause local alterations in the concentration of growth regulatory factors [213]. Support for this idea was sought in a series of further experiments.

Small implants of kidney tissue were grafted into the right lung of immature frogs and when the incisions had healed and the lung reinflated the contralateral lung was removed. This stimulated increased cell division not only in the alveolar tissue of the right lung but also in the parenchymal tissue of the kidney graft [218]. Similar results were obtained using ectopic organ grafts of liver tissue [217].

It is generally accepted that the mitotic response in compensatory growth is strictly organ or tissue specific. The above experiments suggest that it is the location which is specific but that, within this location (the lung in this case), a tissue may respond even if it is of a different type to that which has been damaged or removed. This finding is very difficult to reconcile with any of the main theories of growth regulation, such as wound hormones, chalones or physiological overload, since all these assume a tissue-specific response mediated by tissue-specific chemical growth regulators. Ectopic grafts of kidney and liver were given sufficient time to establish neural and vascular connections with the surrounding lung alveolar tissue and the results are therefore consistent with the idea that cell division is regulated by local physiological changes consequent upon an altered functional state.

This theory presupposes some form of chemically mediated growth regulation but it does not require these mediators to be systemically distributed, since the location of the response is dictated by physiological processes which are known to accompany an altered functional state. As shown above, the need to postulate the existence of tissue-specific regulators stems from the assumption that they are systemically distributed.

The theory seeks to explain by a common mechanism the proliferative response occurring after local tissue damage (acute inflammation), the response to loss or damage to homologous tissue situated elsewhere in the body and the response to altered functional states in the absence of tissue trauma. All three types of response have an important feature in common – namely an altered rate of blood flow due to neurally mediated vascular changes. There is nothing new about the suggestion that cell division may be controlled by alterations in the vascular supply, but the hypothesis has largely been neglected as a result of experimental evidence, the validity of which will now be reviewed.

## IX. The Possible Role of Acute Inflammation and Vascular Changes in the Regulation of Cell Division

### A. Acute Inflammation

The close association between the inflammatory and the proliferative responses has already been noted and the idea of a causal connection is inherent in the concept of wound hormones released from traumatized tissue. Connective tissue and epidermal cell proliferation can be induced by injection of inflammatory exudates and it was suggested that inflammation may play a causal role in stimulating cell division in normal tissues and in inducing the stable changes found in neoplastic tissues [153]. Following manipulation of liver lobes, in the absence of tissue loss or removal, the rate of DNA synthesis increases, possibly due to the local acute inflammatory reaction [246]. There is evidence that inflammation in the absence of tissue damage is not associated with increased cell division but it has been pointed out that the type of damage used in many such experiments severs the normal nerve connections to the tissue and that neural mechanisms may play an important part in the mediation of the proliferative response [8]. An experiment which has been widely accepted as a definitive demonstration that acute inflammation does not promote cell division was performed on the epidermis [40]. A raised flap of skin was wounded on the underside, causing trauma to the hypodermis and dermis, without any effect on the mitotic rate in the overlying epidermis. It was concluded that acute inflammation, including non-specific tissue damage and an increased blood supply, plays no part in inducing cell proliferation. In view of the degree of manipulation involved in this experimental procedure, it is surprising that it was without effect on cell division since even light massage or tension on the intact skin can induce an increase in epidermal mitotic rate [40]. Very light pressure applied to the intact skin in vivo leads to increased premitotic RNA polymerisation in the epidermis, while even very heavy pressures on skin cultures maintained in vitro are without effect and it was suggested that an intact nerve and blood supply are essential [8]. In the experiments on hypodermal wounds [40] the flap of skin was cut on three sides, which would be sufficient to cause considerable disruption of the nervous and vascular supply. These experiments, therefore, may not constitute a valid argument against the role of inflammation and tissue damage.

## B. Changes in Vascular Supply

The relationship between blood flow and mitotic response in the regenerating liver has been studied in some detail [34]. Partial extirpation of the liver approximately doubles the rate of blood flow through the remaining liver, the rate returning to normal after 5 days [23] and it was proposed that this plays a direct role in stimulating cell proliferation. The proliferative response in the regenerating liver is dependent on an efficient vascular supply [245] while the fact that the volume of blood delivered appears to be the determining factor, rather than whether the blood is of portal or caval origin, suggests that the response is due to the flow rate per se rather than to some hypothetical regulatory factor present in the afferent supply [53].

Immediately after birth, the rate of cell division in the liver is low [136] and the rate of growth lags behind that of other organs for several days [167]. Such changes in the rate of cell division at the time of birth may be due to alterations in the hepatic microcirculation which determine the local concentration of extracellular regulatory proteins [137].

Other experimental evidence casts doubt on the existence of a causal relation between increased blood flow and cell division. Anastomosing the splenic artery to the hepatic portal vein increases the blood flow to the liver but has no effect on the rate of cell division [59], and the evidence from other similar experiments is conflicting [207]. The results of experiments on the lung, described above, show that there can be no direct relation between cell division and the total amount of blood supplied to the organ. Ligature of the left pulmonary artery, which is part of the surgical procedure for removing the left lobe, ensures that the whole of the output of the left ventricle is diverted to the right lobes of the lung. However, there were situations where removal or collapse of the left lobe did not give rise to increased cell division in the right lobes. However, if there is any causal relation between increased blood flow and cell division it is likely to be related not so much to the total amount of blood supplied to the organ, but to an increased blood flow in the microcirculation [136] with its intimate connection with tissues and individual cells. As will be seen in the following section, there are many physiological devices by which increased blood flow can bypass the microcirculation.

## X. Control of Blood Flow in the Microcirculation

The microcirculation is defined as that part of the vascular system comprising blood vessels of diameter less than 100 µm. This includes the true capillaries through which the exchange of dissolved gases and other metabolites takes place.

### A. Anatomy of the Vascular Bed

The rate of blood flow through the true capillaries can be regulated by the concerted action of a number of mechanisms [152] which depend on the tone of smooth muscle fibres in the walls of blood vessels. Resistance vessels may be present both in the pre- and post-capillary sections while the flow through individual capillaries may be controlled by precapillary sphincters. Changes in muscle tone in the larger diameter capacitance vessels in the venous section can influence venous return and hence cardiac output. These mechanisms can serve to reduce the total blood supply to an organ, to regulate the relative distribution of blood to different parts of the same organ or to bypass the capillary exchange system through arteriovenous shunts.

### B. Neural Control of Vascular Resistance

Neural control of the tone of smooth muscle, including that in the walls of blood vessels, is exerted through the autonomic nervous system [156] which is characterized by having ganglia, containing synapses, between the central nervous system and the effector organ [126]. The autonomic nervous system comprises the sympathetic system, derived from the thoraco-lumbar outflow of the spinal cord, in which the transmitter substance between the post-ganglionic nerve endings and the effector muscle fibres is noradrenaline, and the parasympathetic system, derived jointly from the cranial (vagus nerve) outflow and the sacral outflow of the spinal cord, in which the neurotransmitter is acetylcholine. Smooth muscle contains two types of receptor, the $\alpha$- and the $\beta$-receptors, capable of responding to sympathetic stimulation [3]. If $\alpha$-receptors preponderate the effect of stimulation is vasoconstriction but if $\beta$-receptors preponderate, as in the vessels of skeletal muscle, the result is relaxation and vasodilatation [156]. The effect of parasympathetic stimulation is vasodilatation [126].

The lack of response following injection of neurotransmitter substances into receptor cells suggests that adrenergic (noradrenaline mediated) and cholinergic (acetylcholine mediated) receptors are situated on the cell surface [237], although there is no general agreement on this point. The chemical nature of the receptor substances is also obscure. β-receptor activation leads to increased cAMP activity, which suggests that the receptor may be part of the adenylate cyclase molecule [120] but this interpretation seems unlikely in view of the fact that the receptor substance and the enzyme are products of separate genes and it has been proposed that the effect of β-receptors is mediated by deinhibition of $Ca^{2+}$-bound enzymes [140]. The action of α-receptors involves an alteration in ion permeability where the first change is a $Ca^{2+}$ influx [140] and there is evidence that this may decrease the cAMP levels [120]. It has been suggested that the α- and β-receptors may be two allosteric forms of the same molecule [140]. Chemical characterisation of the cholinergic receptors has likewise proved elusive, though there is some evidence that it may be a protein or proteolipid [237].

Investigation of the chemical nature of the receptors has been further complicated by recent observations that there may be more than one type of each receptor. On the basis of the proven organ selectivity of different α-antagonistic drugs, it has been proposed that there are at least two types [112] and possibly more [105] of α-receptor [27] while similar methods have led to the identification of more than one type of β-receptor [107, 132, 237].

## C. Regulation of Blood Flow in Specific Organs

There is evidence that the autonomic nervous system is directly involved in the control of the rate of blood flow in a number of different organs.

In the liver there are short anastomoses between the portal and hepatic venules which can act as shunts, diverting blood from the long branching sinusoids [128]. It has been suggested that the local regulation of blood flow through the hepatic sinusoids is mediated by vasodilator metabolites, such as adenosine, released from hypoxic cells following rapid glycogenesis [150] but more recent evidence suggests that the vasomotor tone, and hence the rate of blood flow, in the sinusoids is under direct autonomic control [129, 146], probably involving the action of the sympathetic system [179].

The gastric circulation appears to be under the control of the sympathetic system with β-receptors stimulating vasodilatation and α-receptors causing vasoconstriction [264]. In the lung there are arterio-venous shunts which can divert blood away from the alveolar microcirculation [66] and it appears that local circulation in pulmonary tissue is controlled by the autonomic nervous system [72, 231]. The sympathetic innervation of the lung has both vasoconstrictor and vasodilator actions, while the parasympathetic innervation has a vasodilator effect [157].

The rate of blood flow in the kidney is probably determined by smooth muscle tone but surgical and pharmacological denervation was found to have no effect on blood flow and it was concluded that autoregulation of blood flow in this organ is not dependent on neurogenic control [235] but that it is controlled by changes in the level of circulating renin [122]. Renin is produced in the juxtaglomerular cells of the kidney, probably as a result of the action of plasma kallikrein on prorenin [191]. It has an enzymatic effect, converting an $\alpha$-globulin renin substrate into two vasoactive substances, angiotensin I, which has a specific vasoconstrictive action on the inner cortex and the medulla and angiotensin II, which acts on the outer cortex [170]. However, autoregulation of blood flow in the kidney is not abolished in renin-depleted animals and it was concluded that the renin-angiotensin system is not a major factor in the control of local blood flow in this organ [91]. Stimulation of sympathetic fibres in the kidney leads to vasoconstriction [102], probably through the action of $\alpha$-adrenergic receptors [30].

The cerebral circulation is under the control of sympathetic nerve action which activates $\alpha$-adrenergic receptors resulting in vasoconstriction [143]. In the cerebral surface, the fine control of the microcirculation appears to occur through action on the precapillary arterioles and precapillary sphincters [29].

Denervation of the skin causes relaxation of large arterial vessels, abolishing their ability to contract spontaneously, but after denervation the small vessels decrease in diameter and show enhanced contractile ability, possibly due to control by local axon impulses [249].

*XI. Chemical Mediation of the Proposed Neural and Vascular Effects*

*A. Blood Flow and the Exchange of Metabolites*

Intermittent activity of the precapillary sphincters, under autonomic nervous control, causes transient fugacity gradients of water extending from the active to the closed capillaries. The flow of water along these gradients is responsible for the transport of metabolites to and from the parenchymal cells of the tissue concerned [12]. The autonomic innervation also controls the permeability of microvessels and of plasma membranes [79]. In this way, large molecules, including γ-globulins, can be carried from the extravascular spaces into the capillaries [171]. The passage of small molecules through the vascular wall occurs uniformly throughout the microcirculation but the movement of large molecules is restricted to capillaries and venules [163], becoming more rapid with increased venous pressure [89]. In addition to the microcirculation, the lymphatic system also contributes towards the removal of soluble materials from the tissues [265]. It has been suggested that the increased DNA synthesis which follows from increasing the blood supply to a tissue is due to alterations in the local pool size of regulatory molecules [141].

## B. Chemical Mediators of Acute Inflammation and Their Effect on Cell Division

The mediation of acute inflammation has been discussed above and many of the chemical substances concerned have been shown to have an effect on cell division.

A large number of vasoactive substances, including histamine [238], thrombin and whole platelets, ADP, serotonin and noradrenaline [67] stimulate increased proliferation of cells, probably by causing an influx of $Ca^{2+}$ ions. Thrombin stimulates the proliferation of chick embryo cells, the effect probably being mediated by a thrombin-sensitive cell surface protein [233]. Platelets and platelet components enhance the replication of cultured endothelial cells and this effect may be due to the release of ADP and serotonin [196]. Thrombin and platelet factors also stimulate proliferation of fibroblasts, HeLa cells, monkey kidney cells [174] and smooth muscle cells [193]. The mitogenic response of a number of cultured lines of fibroblasts was found to be due to the potentiating action of thrombin on other growth-promoting substances such as prostaglandins ($PGF_{2\alpha}$), epidermal growth factor and fibroblast growth factor [262].

Many types of prostaglandin also have an effect on cell division. $PGA_1$ and $PGE_1$ were found to enhance proliferation of thymic lymphocytes by changing the intracellular cAMP level, though $PGF_1$ and $PGF_{2\alpha}$ were without effect [88]. The stimulatory effect of $PGE_1$ on myoblasts [261] and of $PGE_2$ on thymic lymphocytes was found to be mediated by the cAMP system but its effect on bone marrow cells appeared to be mediated in a different fashion [82]. Low concentrations of $PGF_{2\alpha}$ increased DNA synthesis in cultured fibroblasts through an increase in intracellular cGMP and a decrease in cAMP [71]. $PGF_{2\alpha}$ also promoted cell division in the intestinal epithelium [238]. Prolactin increased the rate of RNA synthesis in cultured mammary cells by enhancing cGMP and decreasing cAMP synthesis. This effect was found to be mediated by $PGF_{2\alpha}$ while $PGA_2$ and $PGA_1$ stimulated cAMP production and abolished the effect of prolactin [186]. It is possible that prostaglandins may also cause the enhanced proliferation of monocytes observed after tissue injury [103].

Many of the substances involved in neurotransmission have been shown to have a direct effect on cell division. β-adrenergic agents increase the level of epidermal cAMP and may thereby modify the rate of cell division in this tissue [175]. α-adrenergic blocking drugs inhibit mitosis in the regenerating liver while β-blockers stimulate mitosis [243]. This may be a direct effect but it is also possible that the action may be mediated via the control of vasomotor tone. Noradrenaline reduces the level of cAMP in thyroid cells by stimulation of α-adrenergic receptors [258] while in the same tissue, stimulation of the β-adrenergic receptors leads to an increase in cAMP [223]. One of the effects of these altered cAMP levels is likely to be a change in the rate of cell proliferation.

The autonomic nervous system appears to play a particularly important role in the control of cell proliferation in the intestinal epithelium. Blockage of sympathetic activity using guanethidine decreased the rate of cell division in crypt cells of the ileum and depressed the circadian rhythm [125]. β-adrenergic stimulation reduced the rate of cell proliferation in the intestinal epithelium [238], prolonged the cell cycle time and decreased the proliferating cell population in the jejunal epithelium [204]. α-adrenergic stimulation had the opposite effect in promoting cell proliferation [238].

## XII. Evaluation of Growth Control Hypotheses

Many hypotheses propose some special mechanism for the regulation of cell division and growth. There is a bewildering wealth of such hypotheses, the connection between them is often obscure and much research has been directed towards attempts to validate one particular hypothesis to the exclusion of others. The ideas developed above rather suggest that growth is not a phenomenon involving special control mechanisms which can be studied in isolation from other organismal functions but that it is closely integrated with other physiological processes. Such ideas provide the basis for understanding how many of the existing hypotheses can be incorporated into a larger unified framework.

From the research reviewed above, it seems probable that there are two organ systems particularly involved in growth control, the central and peripheral autonomic nervous system and the blood vascular system. The participation of the nervous system may be at the level of local axon reflexes, initiated by local tissue trauma or it may involve an output from the central nervous system. Activation of central nervous system pathways may occur either in response to sensory information received from tissues or organs via the efferent sensory nerves as a result of trauma or change in physiological function or, in the case of psychosomatic dysfunction, it may be due to stimuli from chronic distress patterns stored in the higher brain centres and relayed by the hypothalamus. Neural effects may be mediated by the direct action of neurotransmitter and receptor substances on cell proliferation or indirectly through the influence of neurotransmitter substances on the tone of smooth muscle fibres in the walls of blood vessels. Vasodilatation, resulting in increased blood flow, is envisaged as stimulating cell division either by causing a local increase in the concentration of substances such as platelet factors which stimulate cell division or by causing diffusion gradients resulting in the local depletion of autoregulative cell products which normally inhibit cell division. The existing theories on the control of cell proliferation, reviewed above, will now be discussed.

### A. Wound Hormones

Many of the chemical mediators of acute inflammation, activated as the result of release of breakdown products from traumatised cells, could influence the rate of cell proliferation. This could occur either through their effect on the local vasculature and the rate of blood flow or more directly, as may be the case with the prostaglandins which influence the cAMP/cGMP system. The fact that the chemical nature of proposed wound hormones has remained obscure is hardly surprising, since it is only in recent years that the substances mediating the inflammatory response have been identified.

### B. Negative Feedback and the Chalone System

As described above, the concept of vascular control embodies the idea of autoregulative inhibitory substances controlling cell division through a process of negative feedback. Since their specificity and site of action are envisaged as being determined by neural and vascular mechanisms there is no a priori need to postulate the existence of substances which, like the putative chalones, are systemically distributed and tissue-specific in action. Studies on intestinal epithelium show that, while cell division is regulated by a systemic feedback mechanism, there is evidence that it is mediated by local rather than humoral factors [185]. However, there is no experimental evidence which concretely disproves the concept of tissue-specific regulators and, in view of the fact that many of the products of differentiated tissues are specific to the cell types concerned, it would not be surprising if autoregulative factors controlling cell division were found to have some degree of tissue specificity.

*C. Physiological Load*

The reason why this theory has found relatively little support is that the search for metabolic substances, whose concentration depends on the degree of functional load and which have a direct effect on cell division, has generally been unsuccessful. It appears more likely that the effect of increased functional load is mediated by neural and vascular mechanisms.

*D. Lymphoid System in Growth Regulation*

There appears to be little evidence for the idea that the rate of cell division is controlled by inhibitory mitotic control proteins elaborated by the lymphoid system. Most existing evidence for lymphoid involvement suggests that the action of lymphoid cells is stimulatory. Evidence has been presented in support of the view that chemical mediators of acute inflammation are implicated in the mechanism controlling cell division. One of the consequences of tissue damage is the formation of new antigens through enzymatic modification of tissue proteins, leading to the antigenic stimulation of lymphocytes which then play a part in regulating the further course of the inflammatory process. The rate of production of antibody-forming cells would thus be expected to influence the rate of cell division, which is in accord with the experimental evidence.

*E. Control of Growth by the Nervous System*

Direct evidence for control by neural mechanisms exists for a number of systems including the salivary gland, regenerating limbs and skeletal muscle. The only cell types or tissues in the body which lack an intimate nerve supply are the migratory blood cells such as erythrocytes, macrophages and neutrophil polymorphonucleocytes. The production of such cells appears to be regulated by specific stimulatory hormones such as erythropoietin and granulopoietin.

## F. Hormones and Other Biological Factors Stimulating Cell Division

It has been proposed that there are three main classes of chemical factors mediating cell division: the agents involved in neurotransmission, the substances which mediate the inflammatory process and locally produced growth inhibitors whose concentration can be modified by the rate of blood flow through the tissue. It is not immediately apparent how the many types of stimulatory growth factor fit into this system, if indeed they play any part at all in the natural regulation of cell proliferation. However, there are good theoretical [248] and experimental [4, 164] reasons for believing that cell division may be controlled by the balance between inhibitory and stimulatory factors and the effect of changes in the rate of blood flow may be to modify the rate of cell proliferation by modifying this state of balance.

Cell proliferation in many tissues, particularly those of the endocrine and secondary sexual organs, is under the control of specific hormones which are transmitted by the blood vascular system. Many such hormones are rapidly broken down and excreted and it therefore seems likely that changes in the rate of blood flow to a target organ would play an important part in determining the availability of the 'trophic' hormones.

## XIII. Control of Cell Proliferation in the Different Types of Growth Phenomenon

The rate at which cells divide appears to be determined by intracellular programmes which are modulated by supracellular regulatory mechanisms in order to maintain ordered development and growth homeostasis. During the earliest stages of embryonic development, the rate of the first cell divisions is probably under the control of cytoplasmic factors since cytoplasmic cleavage may continue even when the zygote nucleus is destroyed or inactivated. These factors include the cytoplasmic deoxyribonucleotides, which form a reserve for the rapid synthesis of DNA, and stable species of mRNA which are laid down in the cytoplasm during oogenesis and which become activated as a result of fertilisation [68]. In differentiating and mature cells, the nucleus comes to play an important part in determining the rate of programmed cell division, probably through stable restriction in the activity of genes controlling the separate pathways of nuclear replication and cytokinesis. The introduction of nuclei from differentiated cells into the cytoplasm of actively dividing cells shows that the expression of genes essential for cell division may be under the control of cytoplasmic factors [108].

During embryonic development, the first indication of the modulation of this intracellular programme by supracellular factors is the induction of the neural tube from neural ectoderm [17], since this involves changes not only in the state of cell differentiation but also in the rate of division. Interaction between cells continues to be an important feature in the growth of many organs, including the lung [5] and gastrointestinal tract [202], and it probably plays an equally important part in the maintenance of growth homeostasis in mature organs [164]. In addition to growth control factors exchanged between different cell types, it also seems probable that cells produce endogenous diffusible factors which regulate the rates of DNA synthesis and mitosis. Much theoretical and experimental evidence supports the view that these may be inhibitory products whose controlling effect is through a process of negative feedback.

In differentiating and mature tissues, the rate of cell proliferation may be determined by the local concentration of the control factors mentioned above. Changes in the rate of blood flow through the microcirculation, by modifying rates of diffusion and clearance of these factors, appear to play an important part in regulating cell division. The selective diversion of blood flow through different parts of the microcirculation is under the control of the autonomic nervous system which, through its central connections, may play a crucial role in coordinating growth and cell division rates at the systemic level. Neurotransmitter substances may also affect cell division directly, without the mediatory action of the microcirculation and the possibility that growth in some tissue may be under direct neural control cannot be excluded.

The joint action of nerves and blood vessels in determining the local concentration of regulatory factors provides a unifying model for understanding many forms of normal growth control, including developmental growth, cell turnover homeostasis and various forms of compensatory processes including wound healing, regeneration and physiological hyperplasia. Disturbance of this regulatory mechanism could provide a basis for the better understanding of various conditions involving disordered growth, such as neoplasia and many forms of pathological hyperplasia, in particular those which have a recognised psychosomatic basis. The possible relevance of neural control mechanisms to the regulation of cell division and neoplasia should be a rewarding field for future research.

## References

1 Abercrombie, M.: Localized formation of new tissue in an adult animal. Symp. Soc. exp. Biol. *11:* 235 (1957).
2 Aherne, W.A.; Camplejohn, R.S.; Wright, N.A.: An introduction to cell population kinetics (Arnold, London 1977).
3 Ahlquist, R.P.: A study of adrenotropic receptors. Am. J. Physiol. *153:* 586 (1948).
4 Alescio, T.; Cassini, A.: Induction in vitro of tracheal buds by pulmonary mesenchyme grafted on tracheal epithelium. J. exp. Zool. *150:* 83 (1962).
5 Alescio, T.; Piperno, C.E.: A quantitative assessment of mesenchymal contribution to epithelial growth rate in mouse embryonic lung developing in vitro. J. Embryol. exp. Morph. *17:* 213–227 (1967).
6 Alfert, M.; Geschwind, I.I.: The development of polysomaty in rat liver. Expl Cell Res. *15:* 230 (1958).
7 Anderson, J.R.: Muir's textbook of pathology; 10th ed. (Arnold, London 1976).
8 Argyris, T.S.: Growth induced by damage. Adv. Morphogen. *7:* 1 (1968).
9 Argyris, T.S.: Chalones and the control of normal regenerative and neoplastic growth of the skin. Am. Zool. *12:* 137 (1972).
10 Argyris, T.S.; Trimble, M.E.: Growth promoting effects of damage in damaged and contralateral kidneys of mouse. Anat. Rec. *150:* 1 (1964).
11 Asano, K.; Amagase, S.: Adenosine-3′, 5′-cyclic monophosphate receptor protein in andrenal cortical mitochondria. Life Sci. *21:* 49 (1977).
12 Ascheim, E.: Traffic of metabolites between blood and tissue. Microvasc. Res. *8:* 64 (1974).
13 Axelrad, A.A.; McLeod, D.L.; Suzuki, S.; Shreeve, M.M.: Regulation of the population size of erythropoietic progenitor cells. Cold Spring Harb. Conf. Cell Prolif. *5:* 155 (1978).
14 Babaeva, A.G.; Kraskina, N.A.; Liozner, L.D.: Stimulation of mitotic activity in liver cells of unoperated mice under the influence of lymphoid cells from partially hepatectomized donors. Tsitologiya *11:* 1511 (1969).
15 Babaeva, A.G.; Yudina, N.V.: Repeated amputation of the lower incisors in rats and hypertrophy of the salivary glands and testes. Bull. exp. Biol. Med. *84:* 1181 (1977).
16 Babaeva, A.G.; Yudina, N.V.; Gambarov, S.S.: Genetic differences in reactivity of lymphoid tissue during regeneration of the liver in different strains of mice. Bull. exp. Biol. Med. *82:* 1091 (1975).
17 Balinsky, B.I.: An introduction to embryology (Saunders, Eastbourne 1975).
18 Barabanov, V.M.; Nikolova-Kitova, D.B.: Bull. exp. Biol. Med. *87:* 349 (1979).
19 Bast, R.E.; Singer, M.; Ilan, J.: Nerve-dependent changes in content of ribosomes, polysomes and nascent peptides in newt-limb regenerates. Dev. Biol. *70:* 13 (1979).
20 Baumgartner, H.R.: The role of blood flow in platelet adhesion, fibrin deposition and formation of mural thrombi. Microvasc. Res. *5:* 167 (1973).
21 Bazerque, P.M.; Meiss, A.; Moroni, M.N.: Autonomous denervation and salivary gland hypertrophy after parathyroidectomy. Acta physiol. latino am. *21:* 107 (1971).
22 Beckingham-Smith, K.; Tata, J.R.: Are new proteins synthesized during hormone induced tadpole tail regression? Expl Cell Res. *100:* 129 (1976).
23 Benacerraf, B.; Bilbey, D.; Biozzi, G.; Halpern, B.N.; Stiffel, C.: The measurement of liver blood flow in partially hepatectomized rats. J. Physiol. *136:* 287 (1957).

24  Bengmark, S.: Liver regeneration. Recent Results Cancer Res. *26:* 187 (1970).
25  Bergeron, M.; Sassier, P.: Evidence for a tissue-specific inhibitor of intestinal cell proliferation in rabbit small intestine muscosa. (Pitman Medical, Tunbridge Wells 1980).
26  Bertalanffy, F.D.; Leblond, C.P.: The continuous renewal of the two types of alveolar cells in the lung of the rat. Anat. Rec. *115:* 515 (1953).
27  Berthelsen, S.; Pettinger, W.A.: A functional basis for classification of α-adrenergic receptors. Life Sci. *21:* 595 (1977).
28  Blanquet, P.R.: Unifying heuristic model of transmembrane coordinate control for cell growth and cell movement. J. theor. Biol. *70:* 345–399 (1978).
29  Blum, B.; Yashin, T.; Benary, W.; Israeli, J.; Davidovich, A.: Responses of the leptomeningeal circulation to hypothalamic stimulation. Microvasc. Res. *13:* 283 (1977).
30  Bomzon, L.; Rosendorff, C.; Scriven, R.L.; Farr, J.: The effect of noradrenaline, adrenergic blocking agents and tyramine on the intrarenal distribution of blood flow in the baboon. Cardiovasc. Res. *9:* 314 (1975).
31  Borle, A.: Cyclic AMP stimulation of calcium efflux from kidney, liver and heart mitochondria. J. Membrane Biol. *16:* 221 (1974).
32  Boynton, A.L.; Whitfield, J.F.; Isaacs, R.J.; Tremblay, R.: The control of human WI. 38 cell proliferation by extracellular calcium and its elimination by SV. 40 virus-induced proliferative transformation. J. Cell Physiol. *92:* 241 (1977).
33  Bradley, W.G.; Jenkison, M.: Abnormalities of peripheral nerves in murine muscular dystrophy. J. neurol. Sci. *18:* 227 (1973).
34  Brauer, R.W.: Liver circulation and function. Physiol. Rev. *43:* 115 (1963).
35  Brugal, G.: Etude autoradiographique de l'influence de la température sur le proliferation cellulaire chez les embryones âges de *Pleurodes waltlii* (Micah).Roux. Arch. EntwMech. Org. *168:* 205 (1971).
36  Brugal, G.: Effects of adult intestine and liver extracts on the mitotic activity of corresponding embryonic tissues of *Pleurodeles waltlii* (Micah). Cell Tiss. Kinet. *6:* 519 (1973).
37  Brugal, G.; Pelmont, J.: Existence of two chalone-like substances in intestinal extract from the adult newt inhibiting embryonic intestinal cell proliferation. Cell Tiss. Kinet. *8:* 171 (1975).
38  Bucher, N.L.R.; Malt, R.A.: Regeneration of liver and kidney. (Little, Brown, Boston 1971).
39  Bullough, W.S.: Mitotic and functional homeostasis: a speculative review. Cancer Res. *26:* 1683 (1965).
40  Bullough, W.S.; Laurence, E.B.: The control of epidermal mitotic activity in the mouse. Proc. R. Soc. *151:* 517 (1960).
41  Bullough, W.S.; Laurence, E.B.: Mitotic control by internal secretion. Expl Cell Res. *33:* 176 (1964).
42  Bünning, E.: The physiological clock (Springer, Berlin 1964).
43  Burch, P.R.J.; Burwell, R.G.: Self and not-self: a clonal induction approach to immunology. Q. Rev. Biol. *40:* 252 (1965).
44  Burgess, A.W.; Metcalf, D.; Russell, S.: Regulation of hematopoietic differentiation and proliferation by colony-stimulating factors. Cold Spring Harb. Conf. Cell Prolif. *5:* 339 (1978).
45  Burton, R.R.; Smith, A.H.: The effect of chronic erythrocytic polycythemia and high altitude upon plasma and blood volumes. Proc. Soc. exp. Biol. Med. *140:* 920 (1972).

46 Bygrave, F.L.: Mitochondrial calcium transport and the regulation of metabolism by calcium in tumor cells; in Criss, Control mechanisms in cancer, p. 411 (Raven Press, New York 1976).
47 Cairnie, A.B.; Lamerton, L.F.; Steel, G.G.: Cell proliferation studies in the intestinal epithelium of the rat. Expl Cell Res. *39:* 539 (1965).
48 Cameron, I.L.: Organismal control of mitotic activity in mice. Expl Cell Res. *32:* 160 (1963).
49 Cannella, H.K.: A difference in the effect of rat liver extract (chalone) on the mitotic index of chick embryo liver when extracted at different times of day. Life Sci. *20:* 155 (1977).
50 Carlson, B.M.: The regeneration of skeletal muscle: a review. Am. J. Anat. *137:* 119 (1973).
51 Carter, B.L.A.; Jagadish, M.N.: The relationship between cell size and cell division in the yeast *Saccharomyces cerevisiae.* Expl Cell Res. *112:* 15 (1978).
52 Chapman, L.F.; Goodell, H.: Participation of the nervous system in the inflammatory reaction. Ann. N.Y. Acad. Sci. *116:* 990 (1964).
53 Child, C.G.; Barr, D.; Holswade, G.R.: Liver regeneration following portacaval transposition in dogs. Ann. Surg. *138:* 600 (1953).
54 Chopra, D.P.; Simnett, J.D.: Changes in mitotic rate during compensatory renal growth in *Xenopus laevis* tadpoles after unilateral pronephrectomy. J. Embryol. exp. Morph. *21:* 539 (1969).
55 Chopra, D.P.; Simnett, J.D.: Stimulation of cell division in pronephros of embryonic grafts following partial nephrectomy in the host *(Xenopus laevis).* J. Embryol. exp. Morph. *24:* 525 (1970).
56 Chopra, D.P.; Simnett, J.D.: Demonstration of an organ-specific mitotic inhibitor in amphibian kidney. Expl cell Res. *58:* 319 (1969).
57 Chopra, D.P.; Simnett, J.D.: Stimulation of cell division in larval kidney *(Xenopus laevis)* by rat kidney antiserum. Expl Cell Res. *64:* 396 (1971).
58 Chung, W.Y.: Calmodulin plays a pivotal role in cellular regulation. Science *207:* 19 (1980).
59 Clarke, A.M.; Thomson, R.Y.; Fraenkel, G.J.: Vascular factors in liver regeneration. Surgery Gynec. Obstet. *126:* 45 (1968).
60 Clayton, R.: Antigens in the developing newt embryo. J. Embryol. exp. Morph. *1:* 25 (1953).
61 Clever, U.: The control of cellular growth and death in the development of an insect; in Rockstein, Baker, Molecular and genetic mechanisms in development and ageing, pp. 33–69 (Academic Press, New York 1972).
62 Cohn, R.: Factors affecting postnatal growth of the lung. Anat. Rec. *75:* 195 (1939).
63 Comfort, A.: The biology of senescence (Churchill/Livingstone, Edinburgh 1979).
64 Comroe, J.H.; Forster, R.E.; Dubois, A.B.; Briscoe, W.A.; Carlsen, E.: The lung: clinical physiology and pulmonary function tests; 2nd ed. (Year Book, Chicago 1962).
65 Cooper, E.H.; Bedford, A.J.; Kenny, T.E.: Cell death in normal and malignant tissues. Adv. Cancer Res. *21:* 59 (1975).
66 Daly, I. deB.; Hebb, C.: Pulmonary and bronchial vascular systems (Arnold, London 1966).
67 D'Amore, P.; Shepro, D.: Stimulation of growth and calcium influx in cultured bovine

aortic endothelial cells by platelets and vasoactive substances. J. cell. Physiol. 92: 177 (1977).
68  Davidson, E.H.: Gene activity in early development (Academic Press, New York 1976).
69  Davidson, J.N.: Wound hormones. Edinb. med. J. 50: 70 (1943).
70  Davies, A.J.S.; Leuchars, E.; Doak, S.M.A.; Cross, A.M.: Regeneration in relation to the lymphoid system. Nature, Lond. 201: 1097 (1961).
71  De Asua, L.J.; Clingan, D.; Rudland, P.S.: Initiation of cell proliferation in cultured mouse fibroblasts by prostaglandin $F_{2\alpha}$. Proc. natn. Acad. Sci. USA 72: 2724 (1975).
72  De Fouw, D.O.; Echt, R.: The influence on pulmonary microvascular perfusion by the autonomic nervous system in intact dogs and cats. Microvasc. Res. 9: 357 (1975).
73  De Lorenzo, R.J.; Freedman, S.D.; Yobe, W.B.; Maurer, S.C.: Stimulation of $Ca^{2+}$-dependent neurotransmitter release and presynaptic nerve terminal protein phosphorylation by calmodulin. Proc. natn. Acad. Sci. USA 76: 1838 (1979).
74  Deuchar, E.M.: Biochemical aspects of amphibian development (Methuen, London 1966).
75  Dewey, D.L.: The identification of a cell culture inhibitor in a tumour extract. Cancer Lett. 4: 77 (1978).
76  Dosne, A.M.; Merville, C.; Drouet, L.; Antonini, G.; Guiffant, G.; Quemada, D.: Importance of transport mechanisms in circulating blood for platelet deposition on arterial subendothelium. Microvasc. Res. 14: 45 (1977).
77  Echave-Llanos, J.M.: Circadian rhythms in DNA, cell production and growth hormone; in Endocrinology. Proc. 4th Int. Congr. Endocrinol., pp. 244–250 (Excerpta Medica, Amsterdam 1973).
78  Ekblom, P.; Miettinen, A.; Saxén, L.: Induction of brush border antigens of the proximal tubule in the developing kidney. Devl Biol. 74: 263 (1980).
79  Eletskii, Y.K.; Astakhova, A.M.; Bykov, A.V.: Effect of vagotomy on permeability processes in structural components of rat liver. Bull. exp. Biol. Med. 87: 293 (1979).
80  Fallon, J.F.; Saunders, J.W.: In vitro analysis of the control of cell death in a zone of prospective necrosis from the chick wing bud. Devl Biol. 18: 553 (1968).
81  Fantes, P.A.; Grant, W.D.; Pritchard, R.H.; Sudberry, P.E.; Wheals, A.E.: The regulation of cell size and the control of mitosis. J. theor. Biol. 50: 213 (1975).
82  Feher, I.; Gidali, J.: Prostaglandin $E_2$ as stimulator of haemopoietic stem cell proliferation. Nature, Lond. 247: 550 (1974).
83  Fisher, B.; Szuch, P.; Levine, M.; Fisher, E.R.: A portal blood factor as the humoral agent in liver regeneration. Science 171: 575 (1971).
84  Fisher, J.M.; Simnett, J.D.: Morphogenetic and proliferative changes in the regenerating lung of the rat. Anat. Rec. 176: 389 (1973).
85  Florey, H.W.: Inflammation; in Florey, General pathology, pp. 21–39 (Lloyd-Luke, London 1962).
86  Floyd, K.; Hick, V.E.; Koley, J.; Morrison, J.F.B.: Effects of bradykinin mediated by autonomic efferent nerves. Q. Jl exp. Physiol. 62: 11 (1977).
87  Folkman, J.: Tumor angiogenesis: role in regulation of tumor growth; in Hay, King, Papaconstantinou, Macromolecules regulating growth and development, pp. 43–52 (Academic Press, New York 1974).
88  Franks, D.J.; MacManus, J.P.; Whitefield, J.F.: The effect of prostaglandins on cyclic AMP production and cell proliferation in thymic lymphocytes. Biochem. biophys. Res. Commun. 44: 1177 (1971).

89  Friedman, J.J.: Transcapillary protein leakage and fluid movement. Effect of venous pressure. Microvasc. Res. *12:* 275 (1976).
90  Frohlich, J.E.; Rachmeler, M.: Effect of adenosine 3'-5'-cyclic monophosphate on cell proliferation. J. Cell. Biol. *55:* 19 (1972).
91  Gagnon, J.A.; Keller, H.I.; Kokotis, W.; Schrier, R.W.: Analysis of role of renin-angiotensin system in autoregulation of glomerular filtration. Am. J. Physiol. *219:* 491 (1970).
92  Glucksmann, A.: Cell deaths in normal vertebrate ontogeny. Biol. Rev. *26:* 59 (1952).
93  Goldbergy, N.D.; Haddox, M.K.; Nicol, S.E.; Acott, T.S.; Glass, D.B.; Zeilig, C.E.: Cyclic GMP and cyclic AMP in biological regulation, in Criss, Control mechanisms in cancer, pp. 99–108 (Raven Press, New York 1976).
94  Goldsmith, H.L.; Marlow, J.C.; Yu, S.K.: The effect of oscillatory flow on the release reaction and aggregation of human platelets. Microvasc. Res. *11:* 335 (1976).
95  Gospodarowicz, D.; Mescher, A.L.: A comparison of the responses of cultured myoblasts and chondrocytes to fibroblast and epidermal growth factors. J. cell. Physiol. *93:* 117 (1977).
96  Gospodarowicz, D.; Moran, J.S.: Growth factors in mammalian cell culture. A. Rev. Biochem. *45:* 531 (1976).
97  Gospodarowicz, D.; Moran, J.S.; Mescher, A.L.: Cellular specificities of fibroblast growth factor and epidermal growth factor; in Papaconstantinou, Rutter, Molecular control of proliferation and differentiation, pp. 33–63 (Academic Press. New York 1978).
98  Goss, R.J.: Adaptive growth. (Logos Press/Academic Press, New York 1964).
99  Goss, R.J.: The physiology of growth (Academic Press, New York 1978).
100  Graham, C.F.: The necessary conditions for gene expression during early mammalian development; in Ruddle, Genetic mechanisms of development, pp. 201–224 (Academic Press, New York 1974).
101  Graham, C.F.; Morgan, R.W.: Changes in the cell cycle during early amphibian development. Devl Biol. *14:* 439 (1966).
102  Green, H.D.; Kepchar, J.H.: Control of peripheral resistance in major systemic vascular beds. Physiol. Rev. *39:* 617 (1959).
103  Greenwood, B.: The stimulation of sheep monocyte mitosis in vitro by autologous plasma taken after surgery and the implantation of foreign material. Q. Jl exp. Physiol. *62:* 107 (1977).
104  Grobstein, C.: The problem of the chemical nature of embryonic inducers; in De Reuck, Knight, Cell differentiation, pp. 131–138. Ciba Fdn Symp. (Churchill, London 1967).
105  Haga, T.; Haga, K.: Characterization of alpha-adrenergic receptor subtypes in rat brain. Life Sci *26:* 211 (1980).
106  Handelman, C.S.; Wells, H.: Morphological and histochemical studies of experimentally enlarged and atrophied salivary glands of rats. Am. J. Anat. *112:* 65 (1963).
107  Harms, H.H.; Zaagsma, J.; De Vente, J.: Differentiation of β-adrenoceptors in right atrium, diaphragm and adipose tissue of the rat. Life Sci. *21:* 123 (1977).
108  Harris, H.: Nucleus and cytoplasm; 2nd ed. (Clarendon Press, Oxford 1970).
109  Harris, M.: Cell culture and somatic variation (Holt, Rinehart & Winston, New York 1964).
110  Herberman, R.B.; Holden, H.T.: Natural killer cells as antitumor effector cells. J. natn. Cancer Inst. *62:* 441 (1979).

111 Hironaka, T.; Miyata, Y.: Transplantation of skeletal muscle in normal and dystrophic mice. Expl Neurol. *47:* 1 (1975).
112 Hirst, G.D.S.; Neild, T.O.: Evidence for two populations of excitatory receptors for noradrenaline on arteriolar smooth muscle. Nature, Lond. *283:* 767 (1980).
113 Holden, E.M.; Janov, A.: in Janov, Holden, Primal man: the new consciousness, pp. 236–257 (Abacus, London 1977).
114 Holliday, R.; Pugh, J.E.: DNA modification mechanisms and gene activity during development. Science *187:* 226 (1975).
115 Horton, E.W.: Prostaglandins at adrenergic nerve endings. Br. med. Bull. *29:* 148 (1973).
116 Houck, J.C.: Circulating factors controlling cell proliferation. Prog. clin. biol. Res. *5:* 193 (1976).
117 Iversen, O.H.: Cybernetic aspects of the cancer problem. Prog. Cybernet. *2:* 76 (1965).
118 Jamieson, G.A.; Vanaman, T.C.; Blum, J.J.: Presence of calmodulin in tetrahymena. Proc. natn. Acad. Sci. USA *76:* 6471 (1979).
119 Janov, A.: in Janov, Holden, Primal man: the new consciousness, pp. 350–366 (Abacus, London 1977).
120 Jenkinson, D.H.: Classification and properties of peripheral adrenergic receptors. Br. med. Bull. *29:* 142 (1973).
121 Johnson, H.A.: in Nowinski, Goss, Compensatory renal hypertrophy, pp. 9–27 (Academic Press, New York 1969).
122 Jose, P.A.; Slotkoff, L.M.; Montgomery, S.; Calcagno, P.L.; Eisner, G.: Autoregulation of renal blood flow in the puppy. Am. J. Physiol. *229:* 983 (1975).
123 Kerr, J.F.R.; Wyllie, A.H.; Currie, A.R.: Apoptosis: a basic biological phenomenon with wide ranging implications in tissue kinetics. Br. J. Cancer *26:* 239 (1972).
124 Kiernan, J.A.: A study of chemically-induced acute inflammation in the skin of the rat. Q. Jl exp. Physiol. *62:* 151 (1977).
125 Klein, R.M.: in Appleton, Sunter, Watson, Influence of guanethidine-induced sympathectomy on crypt-cell proliferation in the pre- and post-closure ileum of the rat (Pitman Medical, Tunbridge Wells 1980).
126 Koizumi, K.; Brooks, C. McC.: The autonomic nervous system and its role in controlling visceral activaton; in Mountcastle, Medical physiology, vol. 2, pp. 783–812 (Mosby, St. Louis 1974).
127 Kolpashikova, IF.; Solopaev, B.P.; Efimov, A.V.; Radkova, O.A.; Gorbunova, L.V.; Mikhailenko, A.A.: Morphological and functional changes in the immune system during reparative regeneration of the liver. Bull. exp. Biol. Med. *84:* 1491 (1977).
128 Koo, A.; Liang, I.Y.S.; Cheng, K.K.: The terminal hepatic microcirculation in the rat. Q. Jl exp. Physiol. *60:* 261 (1975).
129 Koo, A.; Liang, I.Y.S.; Cheng, K.K.: Adrenergic mechanisms in the hepatic microcirculation. Q. Jl exp. Physiol. *62:* 199 (1977).
130 Krokan, H.; Wist, E.; Prydz, H.: Effect of cytosol on DNA synthesis in isolated HeLa cell nuclei. Biochem. biophys. Res. Commun. *75:* 414 (1977).
131 Landreth, G.; Cohen, P.; Shooter, E.M.: $Ca^{2+}$ transmembrane fluxes and nerve growth factor action on a clonal cell line of rat phaeochromocytoma. Nature, Lond. *283:* 202 (1980).
132 Lands, A.M.; Arnold, A.; McAuliff, J.P.; Luduena, F.P.; Brown, T.G.: Differentiation of receptor systems activated by sympathomimetic amines. Nature, Lond. *214:* 597 (1967).

133 Landström, U.; Løvtrup-Rein, H.; Løvtrup, S.: Control of cell division and cell differentiation by deoxynucleotides in the early embryo of *Xenopus aevis*. Cell Differ. *4:* 313 (1975).
134 Laznitski, I.: The action of hormones on cell and organ cultures; in Willmer, Cells and tissues in culture, vol. 1 (Academic Press, London 1965).
135 Leading Article: Hyperthyroidism and Graves' disease Br. med. J. *ii:* 457 (1975).
136 Le Bouton, A.V.: Growth, mitosis and morphogenesis of the simple liver acinus in neonatal rats. Devl Biol. *41:* 22 (1974).
137 Le Bouton, A.V.; Marchand, R.: Changes in the distribution of thymidine $^3$H-labelled cells in the growing liver acinus of neonatal rats. Devl Biol. *23:* 524 (1970).
138 Leffert, H.L.; Weinstein, D.B.: Growth control of differentiated foetal rat hepatocytes in primary monolayer culture. IX. Specific inhibition of DNA synthesis initiation by very low density lipoprotein and possible significance to the problem of liver regeneration. J. Cell Biol. *70:* 20 (1976).
139 Leuschen, M.P.; Amato, R.S.: The role of cyclic nucleotides in the regulation of mitotic activity in SSPE virus-infected human brain tissue. Expl Cell Res. *114:* 79 (1978).
140 Levitzki, A.: Catecholamine receptors. Rev. Physiol. Biochem. Pharmacol. *8:* 1 (1978).
141 Lieberman, I.; Short, J.: Hepatic blood supply and control of DNA synthesis in liver. Am. J. Physiol. *208:* 896 (1965).
142 Lin, J.C.; Baserga, R.: The cell cycle and the neoplastic transformation; in King, Developmental aspects of carcinogenesis and immunity, pp. 65–72 (Academic Press, New York 1974).
143 Lluch, S.; Gomez, B.; Alborch, E.; Urquilla, P.R.: Adrenergic mechanisms in cerebral circulation of the goat. Am. J. Physiol. *228:* 985 (1975).
144 Lockshin, R.A.; Beaulaton, J.: Programmed cell death. Life Sci. *15:* 1549 (1974).
145 Loginov, A.S.; Speranskii, M.D.; Arvin, L.I.; Matyushina, E.D.; Magnitskii, G.S.: Chalones of the liver. Bull. exp. Biol. Med. *82:* 1852 (1976).
146 Lundgren, O.; Jodal, M.: Regional blood flow. A. Rev. Physiol. *37:* 395 (1975).
147 MacDonald, R.A.; Rogers, A.E.; Pechet, G.: Regeneration of the liver: relation of regenerative response to the size of partial hepatectomy. Lab. Invest. *11:* 544 (1962).
148 Mamontov, S.G.: Diurnal rhythm of mitosis in the corneal epithelium after splenectomy. Bull. exp. Biol. Med. *87:* 628 (1979).
149 Marks, F.: Epidermal growth control mechanisms, hyperplasia and tumour promotion in the skin. Cancer Res. *36:* 2636 (1976).
150 McCuskey, R.S.: A dynamic and static study of hepatic arterioles and hepatic sphincters. Am. J. Anat. *119:* 455 (1966).
151 McMinn, R.M.H.; Pritchard, J.J.: Tissue repair (Academic Press, New York 1969).
152 Mellander, S.; Johansson, B.: Control of resistance, exchange and capacitance functions in the peripheral circulation. Pharmac. Rev. *20:* 117 (1968).
153 Menkin, V.: Cellular injury in relation to proliferative and neoplastic response. Cancer Res. *1:* 548 (1941).
154 Menton, D.N.: The effects of essential fatty acid deficiency on the skin of the mouse. Am. J. Anat. *122:* 337 (1968).
155 Millis, A.J.T.; Hoyle, M.; Field, B.: Human fibroblast conditioned media contains growth-promoting activities for low density cells. J. cell. Physiol. *93:* 17 (1977).
156 Milnor, W.R.: Autonomic and peripheral control mechanisms; in Mountcastle, Medical physiology, vol. 2, pp. 944–957 (Mosby, St. Louis 1974).

157  Milnor, W.R.: Pulmonary circulation; in Mountcastle, Medical physiology, vol. 2, pp. 1008–1018 (Mosby, St. Louis 1974).
158  Mizell, M.: Limb regeneration: induction in the newborn opossum. Science *161:* 283 (1968).
159  Mohr, U.; Althoff, J.; Kinzel, V.; Süss, R.; Volm, A.: Melanoma regression induced by chalone. Nature, Lond. *220:* 138 (1968).
160  Morton, G.; Liversage, R.A.: In vitro studies of limb regeneration in adult *Diemictylus viridescens*. J. Embryol. exp. Morph. *33:* 813 (1975).
161  Movat, H.Z.: The kinin system: its relation to blood coagulation, fibrinolysis and the formed elements of the blood. Rev. Physiol. Biochem. Pharmacol. *84:* 143 (1978).
162  Muggleton-Harris, A.L.; Hayflick, L.: Cellular ageing studied by the reconstruction of replicating cells from nuclei and cytoplasms isolated from normal human diploid cells. Expl Cell Res. *103:* 321 (1976).
163  Nakamura, Y.; Wayland, H.: Macromolecular transport in the cat mesentery. Microvasc. Res. *9:* 1 (1975).
164  Needham, A.E.: The growth process in animals (Pitman, London 1964).
165  Nosova, I.M.; Zaidenberg, M.A.; Petrosova, V.N.; Seifulla, R.D.; Karelin, A.A.: Changes in the cyclic nucleotide content in wound tissues during healing. Bull. exp. Biol. Med. *87:* 625 (1979).
166  Oates, C.L.; Simnett, J.D.: Organ growth in relation to cell division and cell size *(Xenopus laevis)*. J. Morph. (in press, 1981).
167  Oliver, I.T.F.; Ballard, J.; Shield, J.; Bentley, P.J.: Liver growth in the early post-partum rat. Devl Biol. *4:* 108 (1962).
168  Otten, J.; Johnson, G.S.; Pastan, I.: Regulation of cell growth by cyclic AMP. J. biol. Chem. *247:* 7082 (1972).
169  Osgood, E.E.: A unifying theory of the etiology of the leukemias, lymphomas and cancers. J. natn. Cancer Inst. *18:* 155 (1957).
170  Peach, M.J.: Renin-angiotensin system: biochemistry and mechanisms of action. Physiol. Rev. *57:* 313 (1977).
171  Perry, M.; Garlick, D.: Transcapillary efflux of gamma globulin in rabbit skeletal muscle. Microvasc. Res. *9:* 119 (1975).
172  Pierce, G.B.: The benign cells of malignant tumors; in King, Developmental aspects of carcinogenesis and immunity, pp. 3–22 (Academic Press, New York 1974).
173  Pitot, H.C.: Neoplasia and differentiation as translational functions; in King, Developmental aspects of carcinogenesis and immunity, pp. 79–88 (Academic Press, New York 1974).
174  Pohjanpelto, P.: Stimulation of DNA synthesis in human fibroblasts by thrombin. J. cell. Physiol. *95:* 189 (1978).
175  Powell, J.A.; Duell, E.A.; Vorhees, J.J.: Beta-adrenergic stimulation of endogenous epidermal cyclic AMP formation. Achs Derm. *104:* 359 (1971).
176  Prader, A.: Testicular growth in puberty; in Berenberg, Puberty, biological and psychosocial components (Stenfert Kroese, Leiden 1975).
177  Prescott, D.M.: Reproduction of eukaryotic cells. (Academic Press, New York 1976).
178  Rafferty, K.A.: Mass culture of amphibian cells; in Mizell, Biology of amphibian tumors, Recent results in cancer research; pp. 52–81 (Springer, Berlin 1969).
179  Rappaport, A.M.: The microcirculatory hepatic unit. Microvasc. Res. *6:* 212 (1973).

180 Rasmussen, H.; Goodman, D.B.P.: Relationships between calcium and cyclic nucleotides in cell activation. Physiol. Rev. 57: 421 (1977).
181 Rathbone, M.P.; Petri, J.; Choo, A.F.; Logan, D.M.; Carlone, R.L.; Foret, J.E.: Noradrenaline and cyclic AMP-independent growth stimulation in newt limb blastemata. Nature, Lond. 283: 387 (1980).
182 Rathbone, M.P.; Vickers, J.D.; Logan, D.M.: Neural regulation of cholinesterase in newt skeletal muscle. J. exp. Zool. 210: 463 (1979).
183 Rebhun, L.I.; Nath, J.; Remillard, S.P.: Sulphydryls, and regulation of cell division. Cold Spring Harb. Conf. Cell prolif. 3: 1343 (1976).
184 Reyer, R.W.; Woolfitt, R.A.; Withersty, L.T.: Stimulation of lens regeneration from the newt dorsal iris when implanted into the blastema of the regenerating limb. Devl Biol. 32: 258 (1973).
185 Rijke, R.P.C.: Some speculations on control mechanisms of cell proliferation in intestinal epithelium; in Appleton, Sunter, Watson, Cell proliferation in the gastrointestinal tract. (Pitman Medical, Tunbridge Wells 1980).
186 Rillema, J.A.: Possible role of prostaglandin $F_{2\alpha}$ in mediating effect of prolactin on RNA synthesis in mammary gland explants of mice. Nature Lond. 253: 466 (1975).
187 Rixon, R.H.; Whitfield, J.F.: Parathyroid hormone and liver regeneration. Proc. Soc. exp. Biol. Med. 146: 926 (1974).
188 Roitt, I.M.: Essential immunology (Blackwell, Oxford 1974).
189 Romanova, L.K.; Leikina, E.M.; Antipova, K.K.: Nucleic acid synthesis and mitotic activity during development of compensatory hypertrophy of the lungs in rats. Bull. exp. Biol. Med. 63: 96 (1967).
190 Rott, N.N.; Shevela, G.A.: Changes in the rate of cell divisions in the course of early development of diploid and haploid loach embryos. J. Embryol. exp. Morphol. 20: 141 (1968).
191 Rumpf, K.W.; Becker, K.; Kreusch, U.; Schmidt, S.; Vetter, R.; Scheler, F.: Evidence for a role of plasma kallikrein in the activation of prorenin. Nature, Lond. 283: 482 (1980).
192 Rushmer, R.F.: The arterial system; in Ruch, Patton, Physiology and biophysics, pp. 600–618 (Saunders, Philadelphia 1965).
193 Rutherford, R.B.; Ross, R.: Platelet factors that stimulate fibroblasts and smooth muscle cells quiescent in plasma serum to proliferate. J. Cell Biol. 69: 196 (1976).
194 Rytomaa, T.; Kiviniemi, K.: Control of granulocyte production. Cell Tiss. Kinet. 1: 329 (1968).
195 Rytomaa, T.; Vilpo, J.A.; Levanto, A.; Jones, W.A.: Effect of granulocytic chalone on acute myeloid leukaemia in man. Lancet i: 771 (1977).
196 Saba, S.R.; Mason, R.G.: Effects of platelets and certain platelet components on growth of cultured human endothelial cells. Thromb. Res. 7: 807 (1975).
197 Saetren, H.: A principle of autoregulation of growth. Expl Cell Res. 11: 229 (1956).
198 Sakai, A.: Humoural factor triggering DNA synthesis after partial hepatectomy in the rat. Nature, Lond. 228: 1186 (1970).
199 Sakai, H.: Contractile properties of protein threads from sea-urchin eggs in relation to cell division. Int. Rev. Cytol. 23: 89 (1968).
200 Sakata, T.; Hikosaka, K.; Shiomura, Y.; Tamate, H.: The stimulatory effect of butyrate on epithelial cell proliferation in the rumen; in Appleton, Sunter, Watson, Cell proliferation in the gastrointestinal tract (Pitman Medical, Tunbrigde Wells 1980).

201 Salmon, S.E.: Immunotherapy of cancer: present status of trials in man. Cancer Res. *37:* 1245 (1977).
202 Sbarbati, R.; Strackee, J.: Aspects of the growth of epithelium and mesenchyme in the intestine of mouse embryos; in Appleton, Sunter, Watson, Cell proliferation in the gastrointestinal tract (Pitman Medical, Tunbridge Wells 1980).
203 Shall, S.; O'Farrell, M.K.; Stone, P.R.; Whish, W.J.D.: Properties of the chromosomal enzyme, poly-ADP-ribose polymerase, in cell growth and replication in normal and cancer cells; in Nakahara, Differentiation and control of malignancy of tumour cells, (University Park Press, Baltimore 1974).
204 Sharp, J.G.; Lipscomb, H.L.; Cullan, G.E.; Fatemi, S.H.; McLaughlin, D.; Crouse, D.A.: Preliminary studies on the effects of hormones on cell proliferation in the gastrointestinal tract of the rat; in Appleton, Sunter, Watson, Cell proliferation in the gastrointestinal tract (Pitman Medical, Tunbridge Wells 1980).
205 Sheldrake, A.R.: The ageing, growth and death of cells. Nature, Lond. *250:* 381 (1974).
206 Short, J.; Brown, R.F.; Husakova, A.; Gilbertson, J.R.; Zemel, R.; Lieberman, I.: Induction of DNA synthesis in the liver of intact animals. J. biol. Chem. *247:* 1757 (1972).
207 Sigel, B.: Extracellular regulation of liver regeneration; in Goss, Regulation of organ and tissue growth, pp. 271–282 (Academic Press, New York 1972).
208 Simnett, J.D.: Reduplication of DNA content in nuclei from organ cultures of mouse lung. Neoplasma *19:* 11 (1972).
209 Simnett, J.D.: Nuclear differentiation in the development of normal and neoplastic tissues; in Sherbet, Neoplasia and cell differentiation, pp. 1–26 (Karger, Basel 1974).
210 Simnett, J.D.: Stimulation of cell division following unilateral collapse of the lung. Anat. Rec. *180:* 681 (1974).
211 Simnett, J.D.; Balls, M.: Cell proliferation in Xenopus tissues. J. Morph. *127:* 363 (1969).
212 Simnett, J.D.; Challoner, D.: Natural changes in the cell division rate in normal tissues and the possible role of adrenalin concentration. Neoplasma *26:* 577 (1980).
213 Simnett, J.D.; Fisher, J.M.: Description of growth phenomena and the formulation of growth control models. Natn. Cancer Inst. Monogr. *38:* 29 (1973).
214 Simnett, J.D.; Fisher, J.M.: Cell division and tissue repair following localized damage to the mammalian lung. J. Morph. *148:* 177 (1976).
215 Simnett, J.D.; Fisher, J.M.; Heppleston, A.G.: Tissue-specific inhibition of lung alveolar cell mitosis in organ culture. Nature, Lond. *223:* 944 (1969).
216 Simnett, J.D.; Heppleston, A.G.: Factors controlling organ growth. Expl Cell Res. *45:* 96 (1966).
217 Simnett, J.; Oates, C.; Walton, J.: Stimulation of cell division in ectopic liver tissue following partial removal of the lung. Experientia *33:* 1457 (1977).
218 Simnett, J.; Walton, J.; Oates, C.: Stimulation of cell division in ectopic kidney grafts following unilateral removal of the lung. Anat. Rec. *187:* 273 (1977).
219 Simpson, S.B.: Regeneration of the lizard tail; in Kiortis, Trampusch, Proc. Regen. in Animals, pp. 431–443 (North-Holland, Amsterdam 1965).
220 Sinclair, D.: Human growth after birth (Oxford University Press, London 1969).
221 Singer, M.: Neutrophic control of limb regeneration in the newt. Ann. N. Y. Acad. Sci. *228:* 308 (1974).

222 Sorokin, S.: Recent work on developing lungs; in De Haan, Ursprung, Organogenesis, pp. 467–492 (Holt, Rinehart & Winston, New York 1965).
223 Spaulding, S.W.; Burrow, G.H.: β-Adrenergic stimulation of cyclic AMP and protein kinase activity in the thyroid. Nature, Lond. *254:* 347 (1965).
224 Spector, W.G.; Willoughby, D.A.: Vasoactive amines in acute inflammation. Ann. N. Y. Acad. Sci. *116:* 839 (1964).
225 Strehler, B.: Time, cells and aging (Academic Press, New York 1977).
226 Sunter, J.P.: Experimental carcinogenesis and cancer; in Appleton, Sunter, Watson, Cell proliferation in the gastrointestinal tract (Pitman Medical, Tunbridge Wells 1980).
227 Surani, M.A.H.; Barton, S.C.; Burling, A.: Differentiation of 2-cell and 8-cell mouse embryos arrested by cytoskeletal inhibitors. Expl Cell Res. *125:* 275 (1980).
228 Susskind, W.: Psychosomatic aspects of skin disease; in Munro, Psychosomatic medicine, pp. 8–16 (Churchill-Livingstone, Edinburgh 1973).
229 Swann, M.M.: The control of cell division. I. General mechanisms. Cancer Res. *17:* 727 (1957).
230 Swann, M.M.: The control of cell division. II. Special mechanisms. Cancer Res. *18:* 1118 (1958).
231 Szidon, J.P.; Fishman, A.P.: Autonomic control of the pulmonary circulation; in Fishman, Heckt, The pulmonary circulation and interstitial spaces, pp. 239–265 (University Press, Chicago 1969).
232 Teir, H.; Lahtiharju, A.; Alho, A.; Forsell, K.J.: Autoregulation of growth by tissue breakdown products; in Teir, Rytomaa, Control of cellular growth in adult organisms, pp. 67–82 (Academic Press, London 1967).
233 Teng, N.N.H.; Chen, L.B.: Thrombin-sensitive surface protein of cultured chick embryo cells. Nature, Lond. *259:* 578 (1976).
234 Thornley, A.L.; Laurence, E.B.: The specificity of epidermal chalone action: the results of in vivo experimentation with two purified skin extracts. Devl Biol. *51:* 10 (1976).
235 Thurau, K.: Renal haemodynamics. Am. J. Med. *36:* 687 (1964).
236 Trampusch, H.A.L.; Harrebomée, A.E.: Dedifferentiation a prequisite of regeneration; in Kiortsis, Trampusch, Proc. Regen. in Animals, pp. 341–376 (North-Holland, Amsterdam, 1965).
237 Triggle, D.J.; Triggle, C.R.: Chemical pharmacology of the synapse (Academic Press, London 1976).
238 Tutton, P.J.M.; Barkla, D.H.: A final common pathway promoting cell proliferation in normal and neoplastic intestinal epithelia; in Appleton, Sunter, Watson, Cell proliferation in the gastrointestinal tract (Pitman Medical, Tunbridge Wells 1980).
239 Tsanev, R.; Sendov, B.: A model of the regulatory mechanism of cellular multiplication. J. theor. Biol. *12:* 327 (1966).
240 Tsanev, R.; Sendov, B.: A model of cancer studied by a computer. J. theor. Biol. *23:* 124 (1969).
241 Van Scott, E.J.; Ekel, T.M.: Kinetics of hyperplasia in psoriasis. Archs Derm. *88:* 373 (1963).
242 Voorhees, J.J.; Duell, E.A.; Bass, L.J.; Harrell, E.R.: Role of cyclic AMP in the control of epidermal cell growth and differentiation. Natn. Cancer Inst. Monogr. *38:* 47 (1973).
243 Vunder, P.A.; Vunder, V.P.; Andronova, T.A.: Effect of adrenergic receptor blocking agents on mitotic activity of the regenerating liver. Bull. exp. Biol. Med. *82:* 1718 (1976).

244 Walter, J.B.; Israel, M.S.: General pathology; 5th ed (Churchill/Livingstone, Edinburgh 1979).
245 Weinbren, K.: The portal blood supply and regeneration of the rat liver. Br. J. exp. Path. 36: 583 (1955).
246 Weinbren, K.; Arden, G.H.; Sterling, G.A.: Increased incorporation of $^3$H thymidine into rat liver DNA after handling of liver lobes. Br. J. exp. Path. 50: 540 (1969).
247 Weiner, H.: Emotional factors; in Werner, Ingbar, The thyroid; 4th ed., pp. 627–632 (Harper & Row, New York 1978).
248 Weiss, P.; Kavanau, J.L.: A model of growth and growth control in mathematical terms. J. gen. Physiol. 41: 1 (1957).
249 Wiedeman, M.P.: Blood flow through terminal arterial vessels after denervation of the bat wing. Circulation Res. 22: 83 (1968).
250 Wigglesworth, V.B.: The principles of insect physiology (Champman & Hall, London 1972).
251 Wilhelm, D.L.: Chemical mediators; in Zweifach, Grant, McCluskey, The inflammatory process, Vol. 2, pp. 251–301 (Academic Press, New York 1973).
252 Williams, T.J.; Peck, M.J.: Role of prostaglandin-mediated vasodilatation in inflammation. Nature, Lond. 27: 530 (1977).
253 Williamson, R.C.N.; Malt, R.A.: Relative importance of luminal and systemic factors in the control of intestinal epithelium; in Appleton, Sunter, Watson, Cell proliferation in the gastrointestinal tract (Pitman Medical, Tunbridge Wells 1980).
254 Willoughby, D.A.: Mediation of increased vascular permeability in inflammation; in Zweifach, Grant, McCluskey, The inflammatory process, vol. 2, pp. 303–331 (Academic Press, New York 1973).
255 Wintersberger, E.: DNA replication in eukaryotes. Rev. Physiol. Biochem. Pharmacol. 84: 93 (1978).
256 Wolpert, L.; Lewis, J.; Summerbell, D.: Morphogenesis of the vertebrate limb; in Cell patterning. Ciba Fdn Symp. No. 29, pp. 95–130 (1975).
257 Wright, N.A.: Cell proliferation in normal experimental animals; in Appleton, Sunter, Watson, Cell proliferation in the gastrointestinal tract (Pitman Medical, Tunbridge Wells 1980).
258 Yamashita, K.; Yamashita, S.; Ogata, E.: Regulation of cyclic AMP levels in canine thyroid slices. Life Sci. 21: 607 (1977).
259 Yasuda, H.; Hanai, N.; Kurata, M.; Yamada, M.: Cyclic GMP metabolism in relation to the regulation of cell growth in Balb/c 3TC cells. Expl Cell Res. 114: 111 (1978).
260 Yen, A.; Fried, J.; Kitahar, T.; Strife, A.; Clarkson, B.D.: The kinetic significance of cell size. I. Variation in cell cycle parameters with cell size measured at mitosis. Expl Cell Res. 95: 295 (1975); II. Size distribution of resting and proliferating cells during interphase. Expl Cell Res. 95: 303 (1975).
261 Zalin, R.: The cell cycle, myoblast differentiation and prostaglandin as a developmental signal. Devl Biol. 71: 274 (1979).
262 Zetter, B.R.; Sun, T.T.; Chen, L.B.; Buchanan, J.M.: Thrombin potentiates the mitogenic response of cultured fibroblasts to serum and other growth promoting agents. J. cell. Physiol. 92: 233 (1977).
263 Zinzar, S.N.; Karmanova, N.V.; Svet-Moldavskii, G.Y.: Inhibition of growth of embryonic tissues transplanted into syngeneic newborn recipients. Bull. exp. Biol. Med. 81: 595 (1976).

264 Zinner, M.J.; Kerr, J.C.; Reynold, D.G.: Adrenergic mechanisms in canine gastric circulation. Am. J. Physiol. *229:* 977 (1975).
265 Zweifach, B.W.: Functional behaviour of the microcirculation (Thomas, Springfield 1961).

J.D. Simnett, MA, BSc, PhD, University of Newcastle upon Tyne,
Department of Pathology, Royal Victoria Infirmary,
Queen Victoria Road, Newcastle upon Tyne NE1 4LP (England)

# Growth Differentiation and Function of Tumours in Organ Culture

*Gisèle M. Hodges*

Imperial Cancer Research Fund, London, England

## I. Introduction

As variously discussed, tumour development may follow cellular alteration which engenders escape from the regulatory stimuli of the organism, and the neoplastic mass of tissue so generated possesses a capacity for relatively autonomous growth [27, 76, 194]. In a majority of the solid neoplasms an organoid character is expressed and within these neoplasms may be recognized various tissue components – populations of neoplastic cells, the surrounding interstitial tissue and a supporting vascular system [71, 76, 86, 194]. These tissues show a spatial arrangement forming, in general, a structured relationship one to another the complexities of which may vary and be associated with other independently variable and dissociable behavioural characteristics of the neoplasm.

In various combinations, these behavioural characteristics can provide a relative distinction between benign and malignant neoplasms, notwithstanding the phenomenon of tumour progression as emphasized by *Foulds* [76]. Neoplasms of the benign type are usually highly differentiated, slow growing, encapsulated, differing only minimally from their normal precursors, and noninvasive. By contrast, neoplasms of the malignant type ordinarily consist of rapidly growing cells that may have a variety of abnormal chromosomes, usually are pleomorphic with various degrees of anaplasia and spread from their site of origin demonstrating a capacity for metastasis [191].

Tumours are complex, heterogeneous entities not only in their assembly of varied tissue components but also in the considerable biological heterogeneity manifested by the constituent neoplastic cell populations. This has been exemplified with respect to variation in their morphological, proliferative, karyotypic, antigenic, metastatic, biochemical, hormone receptor and immunogenicity properties, and by the different behavioural responses expressed by neoplastic tissues to radiation or to cytostatic drugs [45–47, 187, 191, 194, 228]. Thus, in establishing strategies to examine the biology and control of tumours, cognizance of the cellular and tissue heterogeneity within neoplasms and of their programmed interactions is imposed [46, 86, 140]. It has long been appreciated that basic information examining many aspects of tumour behaviour cannot be obtained solely from in vivo studies but depends on the development of appropriate in vitro models. Tissue culture has clearly broadened the approach for an understanding of tumour biology based largely on the systems of primary or established cell culture. Nevertheless, on the basis of the earlier discussion, there are compelling reasons for applying in vitro models in which retention of the organoid structure and function of neoplasms is favoured. Such systems may have the possible advantage that a relatively intact epithelial-stromal relation could effect more representative reactions than those obtained with other techniques and provide basic information in areas such as tumour responsiveness to hormones or to chemotherapeutic agents, tumour progression or tumour invasiveness amongst others.

Against this background the review firstly scrutinizes the basic methodology of organotypic culture and certain of those factors which may modify their performance together with a brief survey of various analytical approaches to such systems. In a second section, the effectiveness of organ culture in studies of neoplasms is reviewed in relation to the in vitro maintenance of both morphological and functional characteristics. While, in a third section, the proficiency of this in vitro culture method in investigative studies of tumour behaviour under various experimental conditions is briefly considered.

## II. Methodology for Tumour Organ Culture

### A. Organ Culture Systems

A variety of tissue culture techniques have been developed which allow cells, tissues, organ primordia or the whole or parts of an organ to closely reproduce in vitro the three-dimensional patterns of histologic organization and differentiation seen in vivo. Within the organized framework of these in vitro systems cells may divide and retain their function and differentiated state, with non-specific cell proliferation being restricted and eliminated so far as is possible. The terms of 'histiotypic' and 'organotypic' and of 'organ culture' have been used to describe such tissue culture systems in which the preservation of complex architecture and functional intercellular relationships are attempted [214].

Recent reviews of organ culture methodology [51, 103] group organ culture techniques into three broad categories: (a) culture of explants at the surface of solid media; (b) culture of explants at the interface of liquid media, (c) and culture of explants immersed in liquid media. Within each group many methodological variants have been developed and to these should be added ancillary organotypic methods such as cell aggregation [171], direct or indirect tissue recombination [85], and three-dimensional culture using sponge [137, 141]. fibrin foam [119, 120] or perfused hollow fibre [125] matrices. Comprehensive accounts of the basic techniques within each group are to be found in previous reviews of organ culture [12, 51, 103, 170, 183, 247, 268], but their salient features will be considered here.

The concept of explant culture on the surface of solid substrates was first introduced by *Strangeways and Fell* [236], and developed by *Fell and Robison* [62] into the classical technique of organ culture. Plasma-embryo extract clots were used both as nutrient media and support surfaces but the system presented several problems. Notably, the almost invariable liquefaction of the plasma clot in the vicinity of the tissue explants led to non-specific cellular proliferation from the margins of the explants. Frequent transfer of explants to fresh plasma clots after careful cutting away from the cellular outgrowth was imposed both to prevent loss of tissue architecture by uncontrolled cellular outgrowth and to minimize exhaustion of local nutrient levels and accumulation of toxic metabolites. Soft agar (about 1%), supplemented either with serum and/or various tissue extracts, or with defined media, provided a convenient alternative support system and minimized the problems of non-specific cellular growth [231, 274, 278]. A later development, which bridged the solid medium and liquid interface (see below) systems, used agar strips (1–2% agar in a balanced saline solution, and conditioned in medium) as supports in liquid medium-containing culture dishes [209].

The solid medium-agar system of *Wolff* and his co-workers is of particular interest within the context of tumour organ culture methodology, in that it allowed long-term maintenance and growth of several tumours to be fruitfully exploited over many years. Several methodological variants were developed, namely mosaic, double membrane or dialysate systems [224, 274, 275, 279]. In mosaic culture, pieces of embryonic tissue (in general 8½-day chick embryo mesonephros) and of tumour were encased together in strips of vitelline membrane obtained from unincubated chicken eggs. By contrast, in the double membrane culture system, the tumour explants were separated from the mesonephros by a layer of vitelline membrane. Pieces of mesonephros were first placed on the membrane which was folded over the mesonephros; then, pieces of tumour were placed on top of the membrane which was folded a second time to enclose the tumour. In a later development, addition of yeast extract or young chicken liver dialysate was shown to permit the continued growth of established tumour organ cultures in the absence of mesonephros or other embryonic tissues.

These various solid medium-support organ culture systems had however, certain limitations, in particular if plasma-based. These included low diffusion rates of nutrients and of toxic metabolites both to and from the explants; local exhaustion of medium; and restricted sampling and analysis of media limiting the biochemical studies of cultured explants. These limitations led to the development of liquid medium interface systems in which the explants were supported on the surface of the medium by a thin platform. The platform consisted either of a floating 'raft' such as lens paper [29] or a fine-gauge net of cellulose acetate fabric [222]; cellulose acetate filter disks supported on a plastic bar containing holes [5]; or, glass beads [2,

261], small glass cylinders [253], stainless or tantalum steel grids [249] supporting substrates such as perforated cellophane, agar, chorioallantoic membrane, or membrane filter (0.45–8 μm pore size as required) rafts contained in individual or multiwell plastic dishes. At the present time the preferred methods for organ culture would seem to be based on modifications of the Trowell system [249].

A drawback of the stationary liquid medium interface system is that the rates of metabolic and gaseous exchange within the tissue can be expected to vary between those areas at the tissue/gas or at the tissue/medium interface. Several systems for improved medium circulation and gas exchange by continuous flow or automatic replenishment of medium have been devised for organ culture and discussed in detail [63, 70, 74, 113, 129, 162, 172]. The evidence would suggest that such circumfusion systems have considerable potential for organs difficult to maintain by the established static methods.

Porous matrix systems were introduced by *Leighton* [137] in an attempt to improve two-dimensional cell culture studies of normal and tumour cell interaction by providing an environment for the organized, three-dimensional growth of tissue or cell aggregates. The matrix tissue culture approach was developed on the basis that in the body such interactions would be expected to occur in a three-dimensional setting [139, 142, 144] and that the application of two-dimensional culture systems could but provide a limited and restrictive approach. Various matrices have been used including cellulose sponge, either in combination with plasma clot [137] or coated with collagen [141, 143], human fibrin foam [119, 120] or gelatin foam supported on stainless steel grids [188]. Such systems were found to favour both cell growth and migration, and the formation of distinct histological structural units, in particular within the interstices of the matrix.

### B. Factors Influencing Organ Culture

Maintenance of the structural and functional integrity of explanted tissues in vitro imposes the need for careful control of a wide range of culture variables. As variously discussed in earlier reviews [12, 51, 103, 104, 170, 173, 247] and briefly outlined below these include trauma; size, age and heterogeneity of tissue; tissue separation procedures; act of in vitro culture; type of substrate; temperature; gas phase, pH; osmolality and media design.

Careful manipulation of tissues is essential in order to minimize the possible loss of tissue organization which may result from traumatic dissection procedures, and this has led to the application of microsurgical techniques. Viability of human biopsy specimens may be also impaired in the operating theatre by the use of hot surgical instruments, and application of unheated cutting applicances is of importance for purposes of in vitro organ culture. Tissue explants 1–3 mm in thickness are usually prepared, the minimum and maximum size limits being imposed respectively by the inevitable degree of trauma caused by the cutting instrument and the gaseous and metabolic rates of diffusion. Large expanses of tissues can, be explanted, however, and this has the advantage of allowing more of the integral structure of the tissue to be maintained in culture [78, 103, 167]. Such an approach is of particular interest for the in vitro study of tumours where one of the basic difficulties can be the non-homogeneity of the tissue sampled.

In tumours varying morphological and functional attributes in different areas of the same tumours may be encountered; biological heterogeneity of the tumour-cell population; differences in the proportion of tumour cells and of stromal tissue; areas of necrosis; and, non-representative or inert areas either of intercellular substances such as collagen, osteoid or, secretion as in some mucinous adenocarcinomas. As a consequence, explants of tumour tissue may not always be truly representative of the tumour in vivo. A useful approach to the selection of appropriate areas for in vitro culture can be the prior examination of frozen sections prepared from one face of the excised tumour [177, 251]. This potential heterogeneity in sampling means there must be both a rigid control of tissue selection and an application of appropriate cultural and analytical conditions. Together, these must ensure an adequate maintenance and representation of the major components of the neoplastic tissue if in vitro organtypic studies of tumours are to be relevant.

Removal of tissues from a natural in vivo environment to one imposed on them in vitro must inevitably provoke biological changes engendered both by intrinsic cellular metabolism and by exogenous stimuli. In the case of human material, this may be compounded by conditions of transport from the operating theatre to the laboratory, particularly when these are in different parts of a city or country. Selective tissue changes may be engendered therefore, both during the processes of transport and of explant dissection. These are in addition to those which may occur during organ culture and where progressive loss of certain structures and functions or, increased growth and differentiation in other components can be a feature of the explanted tissue. While few comprehensive studies are available on the morphological and metabolic life-history of organtypic cultures as a consequence of the tissue culture environment some useful outlines are to be found in *Auersperg and Finnegan* [11] and in *Balls and Monnickendam* [12]. Implicit in this discussion is the fact that any design and analysis of organ culture systems must take into account the possible selective effect of the in vitro environment on cell and tissue behaviour. Similarly, the introduction of tissue separation procedures into the experimental design will impose various changes at the cellular level. While many of these are transient and may be reversed after a period of time in culture, the selective potential of such procedures on cell populations within tissue explants must be considered as a further variable of the in vitro environment.

The physical nature of the substrate can profoundly influence the in vitro behaviour of tissues, while non-specific cellular proliferation and migration can be restricted by the choice of appropriate substrates [for discussion, see 103]. Of interest, though not widely exploited, is the application of vitelline (and chorionic) membranes of the unincubated chicken egg to encase tissue explants and in particular those such as tumour pieces which do not possess limiting membranes [224, 274, 275, 279]. Membraneous tissues, such as the chick chorioallantoic membrane, the rat omentum, or the chick ammiotic membrane have been used to investigate the invasive character of various cell lines and primary tumours [3, 4] and for which purpose a system of everted intestinal sacs has been also developed [70] (see Section IV, B). Membrane filter substrates of either cellulose acetate or polycarbonate composition have become, in general the preferred organ culture support material. The choice of such supports can influence, however, the propensity of some epithelia to migrate from the explant edge. Epithelium of bladder organ cultures, for example, will establish extensive cellular sheets over cellulose acetate supports though not over polycarbonate supports. Such outgrowths are of interest in that they provide convenient systems for assessing the responsiveness of epithelia to various stimuli under conditions of modified tissue organization.

Tissues are cultured, in general, at temperatures within the body temperature range of the animal from which they are derived. While they may survive and grow at other temperatures, the in vitro behaviour of tissues is usually optimal only within relatively narrow temperature limits. In the knowledge that the complex enzymic and metabolic processes of the body operate each at specific temperature optima, such limits could prove most critical. Mammalian tissues are cultured routinely at 37 °C on the presumption that this temperature is optimal for all cells, but such temperature limits may need to be refined in the design of more specific culture environments. For example, it has been shown that both maintenance and growth of lens, of oral squamous epithelium, or of epidermis are enhanced at temperatures in the region of 32–33 °C [112, 115, 219].

Of considerable importance to the in vitro culture of tissues is the gaseous environment. Most tissue culture media are formulated with a bicarbonate buffer system and the bicarbonate concentration is adjusted so that a 5–10% carbon dioxide gas phase can be used to stabilize the pH of the medium. Physiological requirements of different tissues in vivo may be expected to vary, however, and imposes the need for more precise determinations of optimal nitrogen, oxygen and carbon dioxide levels in vitro [for discussion, see 117, 164, 257]. Furthermore, it is clearly recognized from the work of *Trowell* [249] that explants must be of a size such that oxygen may readily diffuse throughout the explant. While the critical explant thickness will depend on the oxygen consumption of each tissue, explants are usually prepared within a thickness order of 2–3 mm. In large explants a clear correlation has been demonstrated between the gradient of incorporation of radioactive labelled protein precursors and of cell viability, to a $pO_2$ gradient extending from the periphery to the centre of the explant [283].

Optimal pH values for mammalian and avian tissues have been found to be within the range of 6.8–7.8 and most culture media are buffered with phosphates and carbon dioxide/bicarbonate which provides a simple system approximating to some extent to the normal physiological situation in vivo. Other buffer systems such as HEPES (N-2-hydroxyethylpiperazine-N′-2-ethane-sulphonic acid) [223] have been also used to improve pH stability. While there is evidence that the pH of media can influence the in vitro development and differentiation of organs and tissues such studies remain limited. One of the complications in the analysis of pH effects on tissue behaviour is the complex interaction between pH, energy source substrate and carbon dioxide content of the medium [24, 257].

It has been generally assumed that tissue culture media should be iso-osmotic with the blood plasma of the species from which the tissue derives. As discussed by *Waymouth* [256], solutions which are isotonic or iso-osmotic for erythrocytes, are not necessarily appropriate for the survival of other cells within organs or tissues. Cell membrane potentials and the transport of ions and of metabolites can be markedly affected by changes in osmolality and it has been argued that the osmolalities of media should be adjusted in relation to a given tissue system [256, 258, 259].

Most organ culture studies have been relatively short term (days to a few weeks), and in many cases these in vitro time periods are arbitrarily determined by the onset of gross degenerative changes. In part, this may reflect deficiencies in media design. Until recently media prepared from plasma and embryo extract have been used extensively for organ culture. These have been, in general, replaced by the synthetic media of cell culture systems supplemented with serum and various other constituents. Short-term organotypic cultures have been also successfully maintained on completely synthetic media which in most cases are formulated within individual laboratories, the only commercially available synthetic medium designed specifically for organ culture having been for many years *Trowell*'s [249] medium T8. The basic nutritional requirements for tissue culture and the role of serum have been extensively discussed in the literature to which reference should be made for fuller coverage of these areas [103, 122, 244, 257, 259, 260]. The metabolic patterns and the nutritional requirements of organotypic culture systems can be expected to be relatively complex and probably more so than those of cell culture systems. Yet, there is currently little systematic study of organ culture nutrition per se, and still less that of tumour organ culture. Nevertheless, there is convincing evidence that retention of full structural and functional integrity of tissues in vitro is dependent on the presence of particular metabolites, hormones and vitamins at concentrations appropriate to given tissues. There is, therefore, still urgent need for the development of media and particularly of a spectrum of synthetic media designed specifically for individual tissue requirements, and such, that periods of long-term organ culture (weeks to months) may be readily obtained.

## C. *Analysis of Organ Culture*

A considerable amount of organ culture work has been devoted to problems of developmental morphogenesis or of histopathology, and has involved classical, morphological methods of analysis. Development of microchemical techniques over the past 20 years or so has led increasingly to the physiological or biochemical analysis of normal and abnormal functions in many organ systems. While the analytical approach to organ culture is similar to that for any other system, problems can arise which are imposed by the particular characteristics of the organ culture system, as recently discussed by *Franks* [77].

There are several components to an organ culture system – explant, substrate, medium and gaseous content – which may be expected to change throughout an experiment. The explant itself can be also subdivided into three parts, each with a potentially different pattern of behaviour, namely (a) the tissue explant itself (and involving growth and maturation of those cellular components present at the time of explantation); (b) migration and growth of cells from the edges of the epithelial moiety of the tissue over the cut surface of the stroma leading often to a process of epiboly [121] or to regenerative hyperplasia, sometimes of grossly abnormal cells [77, 167], and (c) epithelial cells may appear at the edge of the explant and grow onto the substrate as described for such tissues as bladder [108]; prostate [167]; trachea [91]; skin [121] or vagina [68]. This outgrowth does not in general function normally and may show changes, for example in the maturation process, in function or in response to hormonal stimulation.

As stressed by *Fell* [61] sample explants should always be fixed for histological examination to monitor the state of the tissue. This also provides an effective screen to biochemical experiments where it is essential to establish that biochemical changes are a function of the living tissue and not of in vitro autolysis. The level of tissue viability, the effect of in vitro conditions on morphology, the variation in response both between different areas of the explanted tissue and between explant and outgrowth areas, and the assessment of the effect of different agents on cultures can be all effectively determined by histological study of the tissue. Problems in interpretation may arise, however, from the morphological heterogeneity of the tissue sampled and which may be particularly accentuated in tumour organ cultures as discussed earlier (see also Section III). Such limitations can be minimized, in part, by investigation of representative sections from a large number of explants [130, 201]. Quantitative approaches to these problems can also assist in providing a more objective assessment of tissue behaviour and facilitate statistical analysis [92, 94, 234].

The scanning mode of electron microscopy (SEM) offers a unique and elegant approach for elucidating changes in tissue organization, cell shape, and in cell surface form and function. It has been extensively applied to the morphological analysis of the free natural surfaces of tissues, of cultured cells and, to a lesser extent to that of organ cultures [for discussion see 106, 107]. Attention has begun also to be directed to the internal surfaces not normally exposed in the intact specimen. Furthermore, the introduction of such techniques as immuno-SEM or of X-ray microanalysis is of considerable interest in providing methodology for correlative morphological and functional studies. This type of approach is of value in that morphology and function can be monitored in bulk specimens at continuously variable levels of resolution and allows a more representative sampling of the specimen than can be otherwise obtained by transmission electron microscopy.

Biochemical studies of organ culture may be restricted by the very little amount of tissue available, while that of the medium may be influenced by problems such as rapid metabolism or instability of the substance produced by the tissue explants, products of dead cells, or changes in the medium resulting from incubation at 37 °C. Among the various properties explored have been rate of oxygen consumption, glucose utilization, lactate formation, activity of the Krebs' cycle, dry weight/wet weight ratio, total protein and DNA in addition to the microchemical analysis (including isotope tracer and radioimmunoassay techniques) of other specialized functions [12, 41, 271]. A satisfactory baseline for measurements may be difficult to establish in that such techniques while providing valuable overall quantitative data cannot readily indicate the proportion of cellular and interstitial tissue present nor the growth or function of different cell types within the tissue explant and cannot discriminate between neoplastic and normal cells in tumour explants. As a consequence analytical approaches based on histo-or cytochemistry and/or on autoradiography are also imposed. Evaluation of tissue behaviour in organ culture must include parameters characterizing their proliferation kinetics. Measurements of mitotic incidence using standard metaphase arrest techniques; or, of incorporation rates of radiolabelled specific DNA precursors by liquid scintillation counting or by autoradiographic estimation of the labelling index have been varyingly used in organ culture studies [for discussion see 157, 198, 225] (Section III, B). While the rate of DNA synthesis can be rapidly analyzed by using liquid scintillation counting there is evidence, as discussed by *Simnett and Fisher* [225], that the values obtained do not necessarily correlate to the rate of DNA synthesis calculated from autoradiographic preparations nor to the rate of cell proliferation as determined from the mitotic index (see also Section III, B). Furthermore, DNA synthesis may vary considerably within different regions of the organ culture and has been shown to be extensive in surface areas but greatly restricted within the internal regions [40, 195, 198, 238, 283].

## III. Characteristics of Tumour Organ Culture

### A. In vitro Longevity: Population Survival and Selection

The first attempts at in vitro culture of tumours date back to early studies such as those of *Beebe and Ewing* [18], *Volpino* [254], *Carrel and Burrows* [28] and *Lambert and Hanes* [132], in which cell growth from explants of dog lymphosarcoma, mouse adenocarcinoma, human sarcomas and carcinomas, and Rous sarcoma respectively were reported upon. Since that time a vast literature has accumulated on tissue culture studies of neoplastic tumours, the great majority of which are based on the two-dimensional cell culture system. Organotypic culture systems have been applied, nevertheless, to a wide variety of neoplastic tissues and a survey of this literature is given in the Appendix (but which omits most studies of tissues cultured for 24 h or less) [161].

Cogent arguments in favour of the organotypic culture system for tumour studies have been put forward as briefly discussed earlier and below, and this type of experimental procedure may be closer to the in vivo situation than monolayer cultures for the solid tumours in particular. In monolayer culture only a small and not necessarily representative proportion of the neoplastic tissue may adapt and become established in vitro, notwithstanding that in organ culture the 'representative' sampling of a tumour may be difficult to attain. Another significant argument in the choice of culture system is that expression of various functions such as growth, secretory function or hormonal response can differ according to the system used. In particular, dependence on an intact organization between different tissue components and especially between epithelium and stroma has been implicated, for example, in normal growth and function of cornea [93], breast [133] and prostate [79, 80]; for specific protein synthesis in certain neuronal tumours [226, 227]; for hormone-dependent tissue interactions in human mammary carcinomas [182]; and, for hormonal response of stilboestrol-induced hormone-dependent renal tumours [1, 2]. Significant differences in behaviour pattern can result in monolayer cultures of these various tissues.

As in normal tissues, most organotypic studies of neoplastic tumours have been, with a few exceptions, of relatively short duration (Appendix). Common to many of these tumour organ cultures has been a process of epiboly with proliferation of epithelial cells leading to a complete or partial encapsulation of the explants by either a superficial epithelial monolayer, or epithelial papillary structure. Most explants have shown also a central necrosis of varying extent which can be related to the fibrous character of the explant, tissue density, type of histological structure, and culture period [5, 160, 167, 201, 202, 204, 207, 251].

Several extensive organ culture studies have been carried out encompassing a wide variety of human tumours. Assessments of overall viabilities have given rates of about 65% after 6 days [5, 160], 50% after 12 days [160], or 43% after 11 days [201] of culture, with the viable explants retaining histological characteristics of the original tumours. In studies of specific tumours, *Rousseau* et al. [202] found for human nephroblastoma cultures a viability rate of 60% at 7 days dropping to 35% at 21 days. While, for human gastrointestinal tumours, the highest percentage of viable explants (42.3% at 3 days) and longest periods of maintenance (7–18 days or longer) were obtained with well-differentiated adenocarcinomas [251]. By contrast, moderately differentiated gastrointestinal adenocarcinomas, medullary carcinomas, and fibrous carcinomas showed, statistically significant lower viability rates of 23%, 18.6 and 7%, respectively, at 3 days with a maximum period of maintenance of 6–8 days. Also, viability rates of 20–50% and of 50%, over culture periods of 26 days and of 4–7 days, have been reported for cultures of parathyroid adenomas [60] and of cervical carcinomas [151] respectively. These, and other studies have suggested that malignant tumours characterized by a high level of differentiation are more successfully maintained in organ culture than the poorly differentiated forms [5, 31, 151, 202, 203, 250, 259, 261, 273], while benign tumours can be consistently and effectively maintained in short-term organ culture [31, 134, 135, 151, 167, 263, 266].

In vitro survival of neoplasms may greatly depend, therefore, on their degree of differentiation. But, other factors have been implicated, as discussed for example by *Rousseau* et al. [202], *Heuson* et al. [99], *Tsypkin* et al. [251] and *Franks* [77]. These factors include age of patient and condition of tissue at the time of removal (tumours irradiated before surgery may show greater fragility in organ culture than non-irradiated specimens), while the abundant development of a dense stroma in fibrous neoplasms could be expected to hamper the supply of nutritive substances and oxygen to neoplastic cells and lead to critical accumulation of toxic metabolites in cultured explants. Furthermore, poor development or maintenance of a stromal compartment could also impose an unfavourable effect on the viability of the neoplastic cells and so corroborate the importance of stroma for the growth and function of tissues well established in the literature.

Among the most difficult group of tumours to maintain have been the scirrhous carcinomas [182, 201, 235, 251, 261, 270]. A recent review of organ culture of human breast tumours [182] underlined the difficulty in obtaining successful cultures of human mammary neoplasms by contrast to that of animal mammary carcinomas and related this to the fact that 75% of human breast tumours belong to the scirrhous type. Their poor survival in organ culture was associated with the dense collagen stroma surrounding the neoplastic cells in such tumours. A clear improvement in viability performance was obtained, however, when mammary scirrhous carcinoma explants were subjected to a collagenase pretreatment which resulted in a loosening of this stroma.

Certain histological structures survive better than others in organ cultures of tumours. A feature common to most cultured explants is the early onset of degenerative changes and generalized fibrosis of the connective tissue. The microvasculature of the tumour explant may show endothelial proliferation [160] while hyaline thickening of the capillary blood vessels with sclerosis and obliteration of their lumens can occur as early as 6 days as reported for neuronal tumours [207, 226]. By contrast to the connective tissue stroma, good preservation of nerves and of smooth muscle has been noted [160] remaining intact for as long as 4 weeks [251], though early necrosis of muscle-type cells has been observed in cultures of nephroblastomas [202]. Of the various histological structures, it is the epithelial elements of neoplasms that tend to be maintained and their eventual failure to survive and proliferate in organ culture may reflect inadequacies in media design or relate, as noted earlier, to the loss of stroma [77].

Very few studies have been carried out to determine optimum in vitro conditions for the organ culture of neoplasms and to study variations in the reaction of different types of tissues to these conditions (see Section II). Furthermore, in studies designed to evaluate sensitivity of neoplasms to hormones there should be an obvious interest in using chemically defined media with and without added hormones. Successful short-term maintenance of various carcinomas have been achieved in chemically defined medium, much of this work having centered on mammary carcinomas. These studies have shown that in medium without addition of hormones, explant survival periods can range from around 24 h–4 days [19, 98, 130, 261] to 14 days [99, 181], while mammary fibroadenomas and dysplasias may survive for up to 5–11 days [261] or 20 days [56]. Similar periods of survival have been achieved using chemically defined medium with specific hormonal supplements [19, 95, 98, 99, 130, 181, 182, 246, 261]. Numerous studies have also explored the effect of a variety of hormones on human and animal neoplasms in organ culture using serum-supplemented medium. Insulin has been reported to both improve explant viability and maintain the preculture histological features of normal and neoplastic tissues including human mammary carcinoma [19, 235, 263] and insulin-dependent DMBA-induced rat mammary carcinomas [98]. By contrast, *Wellings and Jentoft* [261] in a study of normal dysplastic, hyperplastic and neoplastic human mammary tissues found no improvement of tissue survival in presence of insulin. Similarly, human scirrhous mammary carcinomas grown in medium supplemented with various combinations of insulin, ovine prolactin and hydrocortisone, showed no improvement in the viability of central tumour cords in these explants [99, 181] whereas further addition of 17β-oestradiol markedly improved tissue survival. Hydrocortisone has also been reported to improve maintenance of tumour organ cultures relative to the controls [246]. Further discussion of neoplastic tissue response to hormones is given in Section IV, A.

Of greater significance for tumour explant viability has been the addition of serum to the medium in concentrations which may range from 5–40% (Appendix) with 10% or less suggested as giving better survival than 20% serum [31] while improved results may be obtained using autologous serum [201]. Serum albumin has also been successfully employed [7]. Also, minimum essential medium has been noted to produce better results than TC 199 when used as the basic element in media design [168, 201].

Of the various tumour organ culture studies, those of *Wolff* and coworkers [277, 279] are of particular interest both with respect to the concept of a continuous organotypic culture of neoplastic tissues, and to their endeavours in the analysis of the in vitro growth requirements of tumour explants. Successful short-term and long-term organ cultures of a variety of mammalian tumours have been attained using the in vitro system developed by these workers, and in which fragments of chick embryo mesonephros tissue provided initially the main source of nutrition. Substances indispensable, under the conditions of the system, for survival and proliferation of the neoplastic cells were shown to be elaborated both by the mesonephros, and to lesser extents by other chick embryo organs (metanephros, liver, gonads, lung and periosteum). It was subsequently shown that these tissue fragments could be substituted by dialysates of mesonephros, chicken liver or yeast extract. Analysis of the in vitro requirements of neoplasms in this organ culture system showed, that in addition to serum and certain essential amino acids (cysteine, methionine, arginine, leucine, and lysine), further substances were essential for the survival and organotypic growth of neoplastic tissues. These growth factors, as elaborated by the chick embryo mesonephros or provided by dialysates of tissue or yeast extracts, were shown to be of low molecular weight, but their further characterization remains to be established.

Tumours, successfully cultured over varying time periods by the Wolff organ culture system, have included mouse sarcoma 180 (157 days); myelosarcoma, uterine and mammary carcinomas of the rat (30 days); mammary adenocarcinoma of the mouse (50 days); rat hepatoma (7 months); and a number of human neoplasias of tissues such as bladder, breast, lung, tongue and uterine cervix; also, various gastrointestinal tissues, lymphoblastic granulomas and lymphomas, and lymph nodes from Hodgkin's disease. The histopathological structure of the parent tumour is, in general, maintained throughout the culture period though, as in other organ culture systems, the connective tissue elements are rapidly lost. Various tumours of the human nervous system (neuroblastomas, ependymoma, medulloblastoma, schwannoma) have been similarly cultured [150]. Pluripotent cells of a mouse teratocarcinoma clonal line have been also successfully transferred to this organotypic system for periods of at least up to 30 days [33]. However, in an earlier study, mouse teratoma cells gave disappointing results with most clones demonstrating poor growth or becoming largely necrotic [59]. Also, a comparative study of tumours grown in presence or in absence of mesonephros fragments failed to demonstrate consistent differences in explant survival between the two groups [160] and is in contrast to the findings of *Wolff* and co-workers discussed earlier.

While many of the tumours cultured by the methodology of *Wolff* and co-workers have shown survival times averaging up to 6 weeks, longer term cultures have been also attained, a number of the neoplasms being subcultured for months or years. In particular, a liver metastasis from a gastric primary tumour (Z 200) has been in continuous culture since 1962; while two colon carcinomas (Z 516 and AZ 110) were maintained for periods of some 10 years [274, 279]; and, a neuroblastoma (E 2) for 1 year [150], these tumour cultures being subdivided as necessary at each transfer of medium. The tumour subcultures may acquire highly characteristic and specific morphologies, explants of Z 200 forming, for example, spherical or ovoid smooth-surfaced nodules, while explants of Z 516 show an irregular outline and deeply corrugated surface. It is useful to reiterate that the initiation of successful cultures was found to depend on the choice of origin of the explant from the tumour and, as discussed by *Wolff* [274], *Wolff* et al. [279] and by *Matoska and Siracký* [159], could be enhanced by using material from the marginal non-necrotic areas of the tumour.

Of the avaible organ culture techniques (see Section II), many have been used for the culture of neoplastic tissues (Appendix) though few attempts have been made to compare differences in their potential as tumour organ culture systems. An exception has been in the study of neoplasms in matrix culture which, as discussed by *Leighton* [139], has various potentialities, although a possible limitation could be its apparently short-term culture possibilities. Yet interestingly, matrix systems can result in successful maintenance of viable cultures for longer time periods than the platform type-membrane filter system as demonstrated in comparative studies of various human and animal neuronal tumours [205, 207, 226, 227] and of a transplantable mouse testicular teratoma [252], with the tumours retaining their original pattern of histoarchitecture in both types of systems during virtually their entire period of viability. Evidence that certain organ culture methodologies may prove more adaptable than others to the long-term organotypic culture of different neoplasms is a further factor to be considered in the development and interpretation of tumour organ culture studies.

As discussed earlier (see Section II) the gaseous environment is of considerable importance to the in vitro culture of tissues but has been little studied in relation to tumour organ cultures. *Rovin* [204], in a study on the influence of $CO_2$ on a variety of neoplasms in vitro, emphasized the importance of appropriate levels of $CO_2$ for explant survival. High oxygen tensions (95%) have been used in the gas phase of the culture environment of a number of tumour organ culture studies (Appendix) yet from the limited work in this area there is little evidence to suggest that such concentrations improve survival or prevent central necrosis [5, 261].

### B. Growth

The problem of evaluating viability and growth of tumours in organ culture has been approached in different ways, mainly on the basis of histologic surveys which, for reasons discussed earlier, can be extremely difficult to quantitate. Comparisons of explant surface area [202] and wet weights of single explants [5], before and after culture, have suggested that no increase in size or weight occurs in organ cultures of normal or neoplastic tissues. Instead, an invariable loss of explant weight would appear to ensue which, after 1 week of culture, can amount to some 10–20% [5]. By contrast, quantitative studies of a human gastric adenocarcinoma (Z 200) have shown marked increases, over a 12-day period, in explant surface area, wet and dry weights (of 300% or more) and total nitrogen (about 400–500%) [156]. Various other neoplasms maintained under the organotypic culture conditions devised by *Wolff* and co-workers [274, 279] (see Sections II and III, A) have also shown active growth over periods of weeks, months or years. These tumour cultures following division into 2, 3 or 4 pieces, grow within 6 days to approximately the same or greater than the dimensions of the previous explant and this process can be maintained continuously over a large number of weekly subcultures. The failure of some tumours to maintain or grow in vitro, other than being a reflection of the tissue culture environment, may relate to the observation that absence of angiogenesis, or prevention of vascularization of solid tumours in vivo, can in some way limit their growth to a small population of limited diameter. This concept emerges from studies on the behaviour of transplanted tumours in isolated perfused organs which provided an insight that a tumour angiogenesis factor might be a control point in tumour growth [71, 72].

Evidence for the viability of neoplastic tissues grown in organ culture may be obtained from various indices of DNA synthesis, viz. specific activity of $^3$H-thymidine or $^{14}$C-thymidine into chemically extracted DNA; number of $^3$H-thymidine-labelled cells; number of mitotic figures; incorporation of $^3$H-uridine or 5-[$^{125}$I]-iodo-2-deoxyuridine [$^{125}$IUdR] into RNA; or, from the nucleic acid or total incorporation of $^{32}$P-phosphate or radioactive amino acids respectively [7, 9, 40, 67, 87, 95, 114, 157, 158, 167, 175, 181, 182, 197, 198, 237, 238, 246, 251, 262, 266].$^{125}$IUdR has advantages over tritium- and $^{14}$C-labelled precursors of DNA in that, as discussed by *Riches* et al. [198], the activity can be assayed on whole organ samples without extensive biochemical and sample preparation, while reutilization of label and retention of unbound activity is minimal. These various indices, used for quantitative measurement of viability of organ cultures and as an index of tumour growth, have been found to provide reasonable quantitative assessments of tumour responsiveness to hormones and drugs in vitro (for further discussion, see Section IV, A, C.).

Quantitative estimations such as $^3$H-thymidine or $^{14}$C-thymidine incorporation into chemically extracted DNA, $^3$H-thymidine labelling indices and mitotic figure analysis can show good direct correlation as demonstrated in organ culture studies of human mammary carcinomas [95, 264, 266], and also correlate well with organ culture evaluations based on histological criteria [158, 181, 212, 246] (but see Section II C).

Procedures for assessing DNA synthesis based on direct measurements of radioactivity in tissues while relatively simple and time-saving cannot take into consideration the morphological and functional heterogeneity of the explants. Autoradiographic studies of mammary [9, 95, 238] and gastrointestinal [251] carcinomas and other neoplasms [40, 195] have shown conspicuous variations in intensity of DNA and RNA synthesis. In particular, nucleic acid synthesis while extensive in marginal areas of the tumour explants has been found to become greatly restricted in the more central areas with the depth of labelling rarely exceeding 100-200 µm [40, 195, 232]. These differences in isotope uptake at different sites of the explant have been attributed both to a decrease of availability of radioactive DNA precursors with tissue depth by slowness of diffusion, and also to differences in oxygen availability [238]. Previous reports have indicated that the depth of the labelled zone depends on the tissue, and will increase if higher $O_2$ partial pressures are applied [195, 232]. These studies have also suggested that while thymidine itself appears to penetrate the tissue well, only those cells which can obtain oxygen by diffusion from the medium actually utilize it [212, 241]. Furthermore, inhomogeneity of labelling in tumour organ cultures may be a reflection of the varying degrees of cellularity and heterogeneous types of cells in the tissue. As noted in studies of human endometrial carcinoma [114], mouse sarcoma 180 [246] and mammary carcinomas [182] this variation in the distribution of labelled cells in replicate samples of the same tissue can be very marked making labelling index determinations difficult to carry out with a reasonable degree of accuracy.

However, as reported in several mammary carcinoma organ culture studies, radioactive DNA precursors are largely incorporated by neoplastic cells with very little labelling being demonstrated by fibroblasts or other non-neoplastic cells [9, 182, 238]. Similarly, mitotic counts have been rarely found in the stroma of such cultures, the majority of cell divisions occurring almost exclusively in the neoplastic cell population of the tumour explants [181]. As a consequence, it has been assumed that in direct measurements of DNA-specific radioactivity the total contribution of non-neoplastic cell labelling to the overall values of radioactive incorporation is insignificant and quantification of the various types of labelled cells or correction of the overall value for this source of error may be considered unnecessary [7, 9]. Replicate samples of the same tissue may, nevertheless, incorporate widely differing amounts of $^3$H-thymidine/mg of tissue which would appear to be a reflection of the differing neoplastic cellular compartments of different areas of the same neoplasm [114, 246]. Such observations provide an indication of the acceptable levels of variation required before definitive hormonal, drug or other effects on neoplastic tissues can be effectively assessed. Other variables shown to affect the quantitative measurement of DNA-specific radioactivity in tumour organ cultures include the amount of tissue (number of explants) and number of replicates assayed; type and amount of medium; labelling time, and culture period [7, 9, 157].

Various short-term changes in the proliferative activity of tumour organ cultures have been reported in the literature. In mammary tumours cultured for 4, 24, 48 and 72 h, a steady decline in the DNA content of explants to about one third of the original value after 72 h has been demonstrated which reasonably mirrors the histological changes and central necrosis in explants over that time period [7, 9]. *McMahon and Thomas* [167] also reported a decrease in total DNA in explants of benign prostatic hyperplasia (BPH) over the first 24 h but which thereafter remained steady, while $^3$H-thymidine incorporation into the DNA was found to increase reaching a maximum activity after 3–4 days culture. Uptake of $^{125}$IUdR measured at 2, 4 and 6 days in cultures of human BPH has shown a similar pattern of activity with the uptake reaching a maximum around day 4 while declining by day 6 to a level similar or lower to that observed during the 1st day of culture [157, 198]. Increased DNA and RNA synthesis, and mitotic activity have been noted in various gastrointestinal tumours after 2–3 days culture with some explants showing a doubling in labelling index values between 24 and 72 h in culture to attain levels of 47% [251]. Incorporation rates of $^{32}$P-phosphate in organ cultures of mouse sarcoma 180 were found by *Tchao* et al. [246] to be always higher at the beginning of the culture up to about 30 h culture compared to a slower rate after 48 h culture. By contrast, labelling index values in organ cultures of endometrial and uterine cervix carcinomas have been shown to decrease over the first 24–48 h or 96 h, and increase at day 4 or day 7 respectively, with the average labelling index values being significantly higher in endometrial carcinomas during the whole period of culture [158, 229]. Similarly, biopsy specimens from patients with fibrocystic diseases of the breast, and from patients with benign and malignant breast tumours maintained in organ culture have also demonstrated differences between pathological groups, though not significantly within the groups in the incorporation rates of radioactive-labelled nucleic acid and protein precursors [67]. Incorporation rates of $^3$H-thymidine, $^3$H-uridine and $^{14}$C-$L$-amino acids were considerably lower in 5-day cultures of fibroadenomas than in cystic mastitis; while in cultures of carcinomas, $^3$H-thymidine incorporation was slightly lower than in cystic mastitis, $^3$H-uridine incorporation was significantly higher while no significant difference was seen with regard to amino acid incorporation.

A feature common to many of these tumour organ culture studies is an increase in the incorporation rates of radioactive-labelled precursors measured during the first 3–4 days in culture followed by a decrease over the next few days. In defining proliferative activity in tumours it is important to consider that such changes may not accurately reflect either the in vivo or the in vitro behaviour of the neoplastic tissue and the results of these studies must be interpreted with great care before they can be applied clinically. Characterization of the tumour growth pattern in organ culture may be influenced, in addition to those factors discussed earlier, by such variables as trauma caused by surgery and subsequent in vitro manipulations, regenerative hyperplasia and tissue culture environment [67, 114, 157, 229] (see also Section II). Periods of in vitro acclimatization may need to be imposed to ensure a stabilization of tissue behaviour though this may be countered by the possible acquisition or loss of intrinsic cellular properties and by the longevity potential of a particular tissue in vitro.

## C. Differentiation and Function

Unique and unusual tissue patterns characterize the histoarchitecture of solid tumours, and these different types of tissue configurations currently provide the basis both of tumour classification and the prognostic parameters of tumour behaviour [76, 194]. An important advantage of organotypic culture systems is that they can retain the grades of histological differentiation of the original neoplasms at least during the initial phases of culture as briefly discussed in a number of studies [23, 118, 119, 139, 158, 207, 208, 251].

In vivo, a variety of the cellular and molecular characteristics of neoplasms may progress in a permanent, irreversible and variably dissociated manner [76, 169, 233]. Progression of the histoarchitecture leads to anaplastic tissue patterns, and which can be associated with the evolution of malignancy and poor prognosis. It has been suggested that organ cultures of neoplastic tissues could serve as simple models for study of tumour progression and in particular, for assessing the malignant potential of tumours, but such approaches remain relatively little explored (see also Section IV, B). In a comparative study of carcinoma and adenomatous polyps of the colon and rectum in fibrin foam matrix cultures, a morphological transition from cytologically benign adenomatous polyps into a focal carcinoma in situ has been suggested in more than two thirds of the culture polyps [118]. Furthermore, some of these cultures also developed invasive features similar to those of colonic adenocarcinoma cultures, these demonstrating a cytologic and histologic architecture comparable to that of the original tumours.

While tumour tissues may retain their original morphologic and histologic type during the initial in vitro stages, stability may be relative and artifactual changes imposed by the act of in vitro culture must not be discounted. Human gastrointestinal adenocarcinomas for example, have been reported as showing in vitro a marked structural simplification of the epithelial tissue after 1–2 weeks of culture leading to unorganized patterns of growth in some of the explants [250, 251]. Alongside such structural changes, a phenomenon of morphologic differentiation not seen in the original tumour tissue may occur as manifested by the onset of epithelial keratinization in cultures of gastrointestinal carcinomas [250] as also in cultures of benign prostate hyperplasia [167] or prostatic adenomas [218] maintained in androgen-free medium.

However, the potentials for histological organization and cellular differentiation of tumours in organ culture have been demonstrated, in particular, in studies of tumours of the nervous system. Progressive gliofibrillogenesis with differentiation to highly fibrillary astrocytes have been reported, for example, in long-term cultures (up to 4 months) of both human benign cerebellar astrocytomas and of anaplastic glioblastomas [207, 226]. The similarity in morphologic expression of differentiation by these two biologically very different types of astrocytic glioma and seen in contrast to the phenomenon of astrocytic glioma dedifferentiation in vivo, was considered indicative of the tendency of organotypic systems to favour differentiation irrespective of the anaplastic character of the original neuronal tumour. Fibrillary differentiation in these cultures was compared to the cellular differentiation known to occur in the normal developing nervous system. As observed in mammalian and avian central nervous cells maintained in vitro, glial filament formation, is a morphologic expression of astrocytic differentiation, and has been demonstrated in these nonneoplastic systems [149]. Organotypic culture systems have also been shown to increase the differentiation not only of human gliomas [206, 207, 226] but of experimental ENU-induced rat astrocytic gliomas [227] and experimental ENU-induced rat malignant schwannomas [205] in contrast to the anaplastic cytological features demonstrated in parallel cultures maintained in monolayer cell systems [207]. In the C-6 rat glioma cell line, which is less differentiated than most known astrocytic gliomas, no increased gliofibrillogenesis was demonstrated in organ culture [21]. However, glial fibrillary acid (GFA) protein, a marker specific for fibrous astrocytes and an indicator of astrocytic differentiation, was shown by immunofluorescence to be markedly increased in organ culture as compared with the same cells grown in monolayer culture. These findings have further supported the view that organ culture systems favour differentiation of glial cells and suggested that, under certain in vitro conditions, the synthesis of GFA protein could proceed in the absence of gliofibrillogenesis; also, that the rat C-6 glioma cell represented a still undifferentiated glial cell type with the potential to differentiate into fibrillary astrocytes.

Neuroepithelial differentiation, similar to that discussed above, has been reported in embryoid bodies of a mouse teratoma transplantable cell line (OTT-6050), though this was dependent on the media design [252]. The embryoid bodies showed, over periods of 7–9 weeks of organ culture, maintenance of the histologic pattern of multiple tissue elements present in the original tumour. Preservation of the neuroepithelial elements was limited to the neuroglial structures only with no maintenance or differentiation of the neuroblastic elements. Organ culture has been also shown to support the successive formation of endodermal vesicles and blood islands in a mouse teratocarcinoma clonal cell line [33].

Experimental studies with respect to the functional, antigenic or karyotypic characteristics and metabolic capacity of tumours in organ culture have yielded data suggesting a relative stability of such properties in vitro, as discussed earlier for tumour histoarchitecture. Gastrointestinal tumours appear to retain the capacity for mucous production both over short-term [250] and long-term [17] culture periods with, in the latter study, secretion of neutral mucopolysaccharides, sulpho- and sialo-mucins being demonstrated in a colon carcinoma (Z 516) after 3½ years of culture. Exposure of human breast infiltrating duct carcinoma explants to the glycoprotein precursors $^3$H-$L$-fucose of $^3$H-glucosamine has provided evidence of the metabolic activity of the neoplastic cells in these short-term cultures (24–48 h), with synthesis of glycoproteins over the Golgi apparatus and the migration of these glycoproteins to other cellular sites and into secretory products [44].

Short-term organ cultures (24 and 48 h) have also been used to examine the activity of various enzymes in MCA-induced mouse carcinomas and in human carcinomas (mammary, ovary and melanoma) in vitro [90]. While succinate dehydrogenase, alkaline phosphatase, adenosine triphosphatase and non-specific esterase showed little activity, NADH-diaphorase, lactate dehydrogenase and acid phosphatase demonstrated activities which were essentially higher than in the original tumour. Quantitative changes in acid phosphatase and β-glucuronidase have also been reported in human prostatic epithelial cells during organ culture [128]. Over short periods of culture (2–4 days) an increase in the number of cells with histochemically demonstrable acid phosphatase occurred but by 8 days these numbers had decreased. In contrast to acid phosphatase activities, the β-glucuronidase activities generally increased during the course of culture, while the ratio of acid phosphatase to β-glucuronidase expressed per cell or milligram of protein decreased with time in culture. Similarly, acid phosphatase activity, while continuing to be located in the differentiated epithelium as in the in vivo tissue, also showed a progressive decrease over a period of 10 days during culture of human BPH [167]. The effects of long-term organ culture on the expression of two surface marker emzymes, alkaline phosphatase and 5'-nucleotidase, have been studied in cultures of N-methyl-N-nitrosurea (MNU)-induced rat bladder tumours [272]. Little 5'-nucleotidase activity was found in the tumour, but extramembraneous alkaline phosphatase, localized as a continuous layer around the cells of the tumour, was still present in the explants after 2 weeks to 2 months in vitro. However, while there was no evidence for loss or decrease in alkaline phosphatase activity, an altered pattern of localization was apparent which took the form of a focal distribution. The relative instability of certain enzymatic functions, as exemplified by these various studies, clearly reflects some deficiency of the in vitro process in these systems.

However, histologic and metabolic studies on pathological human thyroid gland explants (from thyrotoxic glands, colloid goitres, adenomatous nodules and carcinomas) maintained for periods of up to 20 days in a chemically-defined medium, while showing poor maintenance of the carcinoma explants, have demonstrated viable hyperplastic gland explants capable of performing at least some of their normal in vivo metabolic functions including uptake of iodine into the colloid [175]. Also, evaluation of normal and adenomatous parathyroid specimens in organ culture, as measured by hormone production and morphology, has shown that over a culture period of 7–14 days adenoma tissue produced more parathyroid hormone (PTH) than its normal counterpart, with the levels in both groups declining over time [60]. Phytohaemagglutinin stimulation indices of parathyroid adenoma cells were lower, however, compared to control normal human lymphocytes and fell sequentially over a 21-day culture period. Furthermore, somatotrophic adenomas removed from acromegalic patients have been shown to maintain in organ culture their histological differentiation over a 4-week period, and to synthesize and release growth hormone (GH) into the medium without hypothalamic control [185]. The morphology of the cultured tumours remained essentially unchanged but the secretory granules while keeping their initial size disappeared gradually, most of the cells looking chromophobic by the 4th week of culture. The GH concentration in the medium was very high in adenoma cultures (10-fold greater than that from non-tumoral pituitary medium) during the 1st week of culture, remaining high in some cultures up to the 2nd week. Some correlation between the morphological and functional results was observed, GH production being greatest in those explants showing the best histological appearance. In a subsequent study [184] it was shown that somatotrophic adenomas, as well as somatomammotrophic ones could synthesize and secrete prolactin (PRL) in vitro with PRL synthesis from these adenomas increasing during culture over a period of 9–16 days. Yet another example of hormone production in vitro by tumour organ cultures comes from the study of an interstitial tumour of rat testes [192]. Immature ventral prostate explants of rat, when associated with fragments of the tumour, showed minimal proliferation of the connective tissue, maintenance of the prostate epithelium and stimulation of secretory activity. The observations were similar to those obtained when prostate was cultured in medium supplemented with synthetic androgen androstanoline benzoate. In its absence (or in absence of the tumour) the prostatic epithelium was disorganized and the stroma proliferated. Leydig cells composed much of the tumour and were shown to be responsible for the endocrine activity of the tumour on the ventral prostate in organ culture. In a comparable study *Picon* et al. [186] investigated the effects of canine foetal and adult testes and testicular tumours (seminomas, interstitial cell tumours, Sertolinomas) on reproductive tracts from 14½ day old rat foetuses cultured in vitro. While Müllerian inhibiting activity (MIA) was displayed by the foetal canine testes, neither the adult testes nor the testicular tumours (regardless of the origin of the tumour, i.e. Leydig cells, Sertoli cells or germinal cells) were able to produce this MIA factor and to induce regression of the Müllerian duct. Organ cultures of neonatal mouse calvaria labelled with $^{45}$Ca have provided another convenient co-culture bioassay system of application to the detection of osteolytic factors released from neoplastic tissues [55]. Some 60% of human breast carcinomas tested could be shown to possess significant osteolytic activity resulting in considerable losses of calcium from the neonatal bone with the osteolytic effect being to a large extent inhibited by culture in the presence of anti-inflammatory agents.

Human ovarian adenocarcinomas in organ culture have been shown to produce a time-related release of various molecular forms of plasminogen activators [10]. In early cultures, molecular forms of mainly 100,000 daltons were found while in late cultures (up to 4 weeks) lower molecular weight forms of 54,000 and 31,000 daltons were released. By contrast, the production of plasminogen activator was more variable in established ovarian carcinoma cell lines with essentially low molecular forms being released. These observations led to the suggestion that synthesis of the high molecular weight form probably requires participation of the tumour cells in their original histological arrangement.

While few investigations have been undertaken to determine if the antigenic properties of tumours are retained or modified in organ culture, the evidence available would suggest that these properties can persist over long periods of time. Hyperplastic thyroid glands have been shown to retain an organ-specific antigen for 14 days in vitro, whereas cell cultures prepared from the same glands no longer contained the antigen after 3 days [174]. Carcinoembryonic antigen (CEA) has been detected by immunofluorescence both in fresh gastrointestinal tumour tissues, and in long-term organ cultures of colonic (Z 516) and gastric (Z 100) carcinomas after 7 and 8 years of growth in vitro [26]. The antigen appeared to accumulate within the glandular structures of the tumour and this suggested that its release was more difficult in culture than in vivo. Similarly, CEA has been shown by immunoperoxidase staining to be produced and released by human colorectal carcinomas when maintained in both monolayer and organ culture [23]. CEA was demonstrated on the luminal aspects of the neoplastic cells of tumours immediately after surgery and in the corresponding explants after 1 week in organ culture. Radioimmunoassay of the media showed that between 10 and 100 ng CEA/ml was released by tumour cultures into the media, the levels declining with the onset of degenerative changes. The highest levels of CEA were obtained from cultures of well-differentiated highly cellular adenocarcinomas and could be also related to the amount of tissue present in culture.

Little attempt has been made to follow the chromosomal character of tumours in organ culture. In a rare study, *De Grouchy and Wolff* [39] examined the karyotype of a long-term (4½ years) culture of a gastrointestinal carcinoma (Z 200) and showed that the numerical and structural aneuploidy stayed remarkably constant (with minor variations) between cells, between cultures, and between time intervals. A good accord was found between the stability of the tumour structure and the stability of the karyotype.

The importance of the epithelial-stromal complex and the interdependence of these tissue types during embryogenesis has become conclusively demonstrated and the possibility that continued inductive activity occurs in the maintenance of adult tissues and organs confirmed in a limited series of studies [242, 269]. A logical extension has been the concept that neoplasms may represent a breakdown of differentiative control [22, 131] and that normal characteristics might reappear if the pathways whereby they are expressed were to be reestablished. One approach has been to study the effects of inductively active embryonic tissues on tumour explants, on the basis that reexpression of a specific differentiation in the neoplastic cell population, similar to the tissue from which it came, could provide evidence that the basic genetic information for normal growth and differentiation is still present though not being used in the neoplastic tissue.

In vitro association techniques between inductor and reactor (tumour) tissues have been applied to the study of several human and experimental animal tumours [32, 37, 38, 58, 145–148, 203, 220, 221]. Morphological and functional changes approaching that of normal tissues have been demonstrated in some of the studies (table I).

Primitive nephron elements have been shown to develop within 96 h in explants of congenital nephromas grown in direct association with 11-day fetal mouse dorsal brain or spinal cord even in tumour explants grown for 4 days or 2 weeks before tissue association [32]. The capacity of human nephroblastomas, previously cultured for periods of up to 14 days, to differentiate in vitro when cultured for a further 8 days in direct combination with identical volumes of inducer tissues was studied by *Rousseau-Merck* et al. [203]. Improved survival and maintenance of structure was observed in approximately one third of the cultures, whereas an increase in the number of tubules present could be demonstrated in only a limited number of explants. In an earlier study, *Ellison* et al. [58] had reported differentiation in only 1 of 28 explants from transplantable renal tumours associated with rat or mouse embryonic tissues. Although these studies could suggest a specific stimulation of tubule elongation produced by the inducer as observed during renal tubulogenesis, doubt has been expressed. A general conclusion has been that these tissue interactions showed no inductive capacity but were either enhancing the latent ability of the renal tumour to differentiate [58] or exhibiting some nutrient action of the embryonic tissues [203]. The studies of *Wolff* and co-workers [274, 279] have clearly illustrated the marked enhancement of tumour explant survival and growth in organ culture when co-cultured with a variety of embryonic tissues and demonstrated a metabolic cooperation between the

*Table I.* In vitro tumour response to embryonic inductive-type tissues

| Tumour (reactor) tissue | Inductor tissue | Tumour response | Reference(s) |
|---|---|---|---|
| Mouse sarcoma | chick embryo – 5 day notochord | Development of cartilaginous tissue | 220, 221 |
| Mouse mammary carcinoma; mouse sarcoma 180 | chick embryo – primitive streak; 7–9 day mesonephros | tubular organization; slower growth rate on reimplantation into host animal | 147 |
| Human mixed paratiod tumour | chick embryo – primitive streak; 3½ day notochord; mouse embryo – salivary gland | cellular reorganization | 147 |
| Mouse giant cell bone tumour; human osteogenic tumour | chick embryo – 3½ day notochord | reduced number of giant cells; some development of osseous trabeculae | 147 |
| Mouse transplantable renal tumour | chick embryo 3–5 day notochord or spinal cord; mouse or rat embryo 10–13 day dorsal spinal cord | limited tubular differentiation (1 of 28 explants) | 58 |
| Congenital nephroma | mouse embryo – 11-day dorsal brain or spinal cord | primitive nephron elements | 32 |
| Various mouse, rat and human tumours | chick embryo – 8½ day mesonephros | growth; maintenance of some histoarchitecture and functional characteristics | 274, 279 |
| Mouse mammary adenocarcinoma | mouse embryo – mammary mesenchyme | tubular development; reduction in DNA synthesis; changes in nuclear and cytologic morphology; presence of presumptive acid mucopolysaccharide matrix | 37, 38 |
| Human nephroblastoma | chick embryo – 3½ or 5½ day ureter or mesencephalon; mouse embryo – 11–15 day mesencephalon | improved viability and histoarchitecture; some tubular organization | 203 |
| Human mammary carcinoma | mouse embryo – 12 day; mammary or 13 day salivary gland mesenchyme; human fetus – 12–18 week mammary mesenchyme | growth; maintenance of neoplastic cells | 6 |

neoplastic tissue and embryonal cells (see Section II). In a more recent but short-term (10-day) study [6], co-cultivation of neoplastic cells from human mammary scirrhous carcinomas with various embryonic mesenchymes was shown to result in an enhanced growth and maintenance of these cells in the majority of the cultures, whereas no growth occurred in the absence of mesenchyme.

Some evidence of the differentiating effects of embryonic inductive tissues on tumours may be suggested, however, from the studies of *DeCosse* et al. [37, 38]. It was found that an undifferentiated transplantable mammary adenocarcinoma, when grown alone in organ culture for 14 days showed evidence of tubule formation in only 2 of 144 cultures with the others demonstrating a disorganized arrangement of neoplastic cells. When cultured in combination with embryonic mammary mesenchyme either in direct combination or transfilter, several morphological and functional changes in the mammary tumour compatible with differentiation were demonstrated. These included tubule formation ranging from simple to complex branching patterns in 57% of cultures; significant reduction of DNA synthesis; changes in nuclear and cytologic morphology; and appearance of a matrix tentatively identified as containing acid mucopolysaccharides [37]. It was concluded from these observations that changes indicative of organized differentiation in a mouse mammary tumour could be obtained by exposing the tumor to inductively active embryonic tissues in this in vitro model. In a subsequent study [38], an increased number of control and experimental cultures confirmed the earlier observations, while the effects of other types of embryonic mesenchyme (salivary gland, limb), embryonic or postnatal tissues (liver, brain), or formalin- or alcohol-fixed embryonic tissues were compared to that of embryonic mammary mesenchyme. Tissue differentiation was most consistently obtained by direct combination of tumour with embryonic mammary mesenchyme, though embryonic salivary and limb mesenchymes were also capable of promoting some tubule differentiation and mucopolysaccharide formation. It was suggested that these observations most probably represented an inductive effect on the tumour explants in that there was no evidence of differentiation after exposure to post-natal tissues, or to formalin- or alcohol-killed embryonic tissues, though heat-killed mammary mesenchyme could be effective (3 of 21 cultures). Of interest, however, is that control or embryonic tissue-exposed tumours, irrespective of the level of cytodifferentiation attained, when subsequently implanted into mice did not show significant differences in tumour growth rates, survival time of the mice or histology. Such evidence may question the concept of an integral process of tumour redifferentiation, and instead may suggest an independently variable and dissociable process, the potential for redifferentiation varying with different parameters of neoplastic behaviour within individual tumours.

## IV. Experimental Investigations

### A. Tumour Response to Hormones

Tumour responsiveness to hormones has been explored in various in vitro systems, including that of organ culture, in attemps to determine interactions of hormones on individual neoplastic tissues without interference of systemic factors and endocrine homeostatic mechanisms present in vivo. Evidence that tumours may retain in organ culture their histological grade of differentiation, various tissue components, anatomical relationships and function has suggested that the evaluation of hormonal effects on tissues by this type of system may be more related to the in vivo state than by that of monolayer cell culture. Hormone-dependent tissue interactions in human mammary carcinomas [182] and hormonal response of stilboestrol-induced hormone-dependent tumours [1, 2] are retained, for example, only in the organotypic-type systems with significant differences in hormone sensitivity being demonstrated by monolayer cultures of these tissues.

Studies have been designed to evaluate sensitivity of the various constituent tissues of neoplasms to hormones, and to develop predictive assay systems to human material obtained by biopsy or operation both for subsequent hormonal therapy of value in clinical practice and for the screening of new synthetic agents. Many of these in vitro studies have relied on short-term organ culture with morphological, histochemical, and in particular, growth-rate determinations of tissue viability, differentiation and function serving as markers of hormonal effects on neoplastic tissue. Problems in the assessment of the data, residing in such basic difficulties as the possible non-homogeneity of the tumour sampled and defects in culture or media design imposing inappropriate structural or functional changes or responses, have been commented upon in various studies (see Sections II, C, III, B. These and other factors discussed earlier are of important consequence and must be taken into account in the development of significant quantitative predictive assays.

Organ culture has been used by several investigators to study the effects of hormones on endometrial carcinoma [111, 114, 126, 127, 158, 179, 180]. While low concentrations of progesterone and of estrogen have shown, in general, only a marginal effect on tissue viability in short-term cultures of endometrial carcinoma, high progesterone levels usually lead to tissue necroses, this being observed both in neoplastic and normal endometrial cultures [127, 179]. No correlation between endocrine responsiveness of the tumour in organ culture and the degree of tumour differentiation has been noted [158, 179] though carcinomas of different histological grade may possibly vary in their response to different concentrations of progesterone [126]. With increasing progesterone concentrations tissue viability and proliferation of the tumour epithelium in organ culture has been reported to decrease progressively [158, 180] while estradiol, though on its own having no effect, may potentiate the progesterone effect which has appeared to be most marked on the less differentiated tumours [180].

*Hustin* [111] in a study on the effect in vitro of various steroids or peptide hormones on endometrial carcinoma reported a condsiderable reduction in DNA synthesis, as measured by $^3$H-thymidine labelling index, after 4 h incubation of the tumour organ cultures with progesterone or synthetic proestrogens. By contrast, *Jacobelli* et al., [114] on the basis of a more extensive statistical analysis of the data, failed to demonstrate any significant effect on incorporation of radioactive thymidine after a 3 day incubation of endometrial carcinoma cultures with progesterone. This led these authors to question both the sensitivity of the organ culture system and of the analytical approach for detection of small hormonal variations in tumour response.

The hormonal response of benign prostatic hyperplasia and of prostatic carcinoma has been the subject of a number of organ culture studies (Appendix) and reference should be made to the recent reviews of *Lasnitzki* [134, 135] for a comprehensive account and discussion of the data.

Organ culture has proved a powerful method for the study of various aspects of hormonal effects on normal mammary gland development and function and shows evidence of providing a valuable approach to the study of various mammary lesions (hyperplastic, dysplastic and neoplastic). The majority of organ culture studies on tumour response to hormones has centered in particular on the mammary tissue. Interest in the potential of the system with respect to breast cancer resides in part from the difficulty in clinical practice in determining which patients will benefit from hormonal therapy, only one third of the total number responding in general to such treatment. As a consequence a number of attempts have been made to devise simple in vitro tests to predict the therapeutic effects of various hormones on human breast cancer complemented by a range of studies on various rodent models (Appendix). Assessment of mammary tumour response to hormone effects has centered essentially on an analysis of cell metabolism and proliferation based on demonstrable changes in incorporation rates of radioactive-labelled DNA precursors [7–9, 67, 98, 157, 237, 238, 263, 267] (see also Section III, B) while other methods of analysis have included histological grading [261] and the lysosomal fragility test [235].

Several hormones used in studies of mammary tissue development in organ culture have been examined for their effects on mammary tumours. In terms of hormone dependency in vitro mouse mammary tissue has been reported to show a decreasing requirement for estrone, progesterone, cortisol, bovine growth hormone and ovine mammotrophic hormone from normal to preneoplastic to neoplastic tissue [57]. An early study of the levels at which steroid hormones (estrone, 17α-estradiol, progesterone, cortisol, corticosterone, aldosterone and deoxycorticosterone) are toxic to organ cultures of mouse mammary adenocarcinomas provided a comparison of effects and data on minimum toxic doses with progesterone being demonstrated as the most toxic of the steroids used [200].

*Wellings and Jentoft* [261] in an attempt to define the hormonal requirements for in vitro maintenance, growth and function of human dysplastic, fibroadenomatous and carcinomatous mammary tissues found that addition of insulin, aldosterone and ovine mammotropin, alone or in combination, had little effect on survival, cell proliferation or secretory function of the tumour cultures. *Elias and Armstrong* [56] found, however, that at effective concentrations of insulin hyperplastic and squamous metaplastic changes were produced in the epithelium of organ cultures of human mammary dysplasias and fibroadenomas similar to those which insulin produces in normal growth.

Insulin-enhanced DNA synthesis and mitotic activity has been shown to occur in various neoplastic mammary tissues including mouse mammary carcinoma [200]; MCA-induced rat mammary carcinoma [238]; DMBA-induced hormone-dependent rat mammary carcinoma [95–98, 181]; benign human breast tumours [264, 266], and human breast carcinomas [263]. In DMBA-induced rat mammary carcinomas while insulin stimulation of DNA synthesis was found to occur to a considerable but variable extent in the majority of the tumours some proved unresponsive to insulin [95, 97, 98, 181]. The large majority of insulin-dependent tumours were also responsive to the combination of prolactin and progesterone. This combination enhanced DNA synthesis and mitotic activity well above the levels reached in the presence of either hormone alone. By contrast, insulin-independent tumours proved totally insensitive to these hormones. The data from these studies indicated that a majority of the DMBA-induced rat mammary tumours responded to the hormones tested in a manner very similar to the response of the normal rat mammary tissue. An essential difference was the variability of response with a gradual loss of response in some tumours, first to progesterone, then to the progesterone-prolactin combination, and finally to all hormones including insulin. The growth-promoting effect of insulin in explants of DMBA-induced rat mammary carcinomas, accomplished with little or no effect on glucose metabolism, led to the suggestion of a dissociation of the usual tissue culture responses to insulin (viz. stimulation of glucose metabolism and promotion of growth) in these tumour cultures [96].

Prolactin has also been shown to be an effective mitogen in organ cultures of DMBA-induced rat mammary carcinoma [181, 263, 267]. Suppression of this prolactin-induced stimulation of DNA synthesis can result, however, in combination with 17β-estradiol [267] though 17β-estradiol alone showed no significant inhibitory or stimulatory effect on DNA synthesis in cultures of the DMBA-induced rat mammary tumour [7, 267]. Prolactin has proven ineffective on the majority of human mammary carcinoma cultures with only a small proportion (15%) responding to the hormone [263]. More recent studies suggest that prolactin is mitogenic to the epithelium of benign human breast tumours [265] while human placental lactogen (a peptide chemically and physiologically similar to prolactin) has been also shown to be mitogenic to the ductal epithelium of such cultures [261, 266] though human growth factor effects little response [264]. The effect of prolactin on lactalbumin production both by normal and neoplastic human mammary tissue has been the subject of organ culture studies. *Kleinberg* [124] using post-menopausal normal human breast tissue found no response to prolactin whereas stimulation of lactalbumin production was demonstrated by a small number of neoplastic tissues. By contrast, *Wilson* et al. [273] found a stimulation of lactalbumin production in normal but not in neoplastic tissues. It was suggested that this could indicate an absence or deficiency of prolactin receptors varying in different neoplastic tissues, or reflect the effect of different culture conditions in that serum, insulin and hydrocortisone were used in the study of *Kleinberg* [124] whereas no serum or hormones were used by *Wilson* et al. [273].

Studies of the hormone sensitivity of human mammary carcinomas to 17β-estradiol have demonstrated either no significant effect [19, 270] or an apparent dose-response [157] or stimulatory effect on DNA synthesis in some carcinomas [8, 67] this being in contrast to a generally inhibitory effect of the hormone on cultures of fibroadenomas [67]. The metabolism of $^3$H-17β-estradiol with conversion into $^3$H-estrone has been reported to be lower in organ cultures of neoplastic tissues than of fibroadenomatous or non-neoplastic tissues [82]. In an earlier study on the interconversion of $^3$H-17β-estradiol and $^3$H-estrone [270] benign tumours were significantly different from scirrhous carcinomas both in the level of radioactivity incorporation and in the ratios of 17β-estradiol-estrone achieved, with lower levels being demonstrated by the fibroadenomas. Supplementation of the medium with 17β-estradiol in combination with insulin or with insulin and prolactin has been shown to markedly improve survival of scirrhous mammary carcinoma explants [99, 182]. Evidence of collagen digestion around the neoplastic cells led to the implication of an estrogen-dependent collagenolytic enzyme system in the collagen-rich mammary tumours.

Testosterone has been found to impose some stimulatory effect on organ cultures of DMBA-induced rat mammary carcinomas [7] and of some human mammary carcinomas [67] while proving either inhibitory or ineffective to other cultures of benign or neoplastic human breast tumours [8, 19, 67]. In a study on the metabolism of testosterone-4-$^{14}$C by non-neoplastic, fibroadenomatous and neoplastic human breast tissue in organ culture, the retention, uptake and metabolism of the hormone was reported to be invariably lower in the neoplastic tissue and still lower in the fibroadenomas [83].

## B. Tumour Invasiveness

Relatively little is known about the mechanism(s) of tumour invasion despite its obvious importance in the metastatic process which involves a series of sequential steps resulting in the establishment of distant tumour foci away from the primary lesion [52, 64–66, 191, 243]. Neoplastic cells endowed with the property to produce metastases are characterized by their ability to invade local host tissue, to penetrate blood vessels and lymphatics and also organ and body cavities, to arrest and adhere to serosal, mucosal or endothelial surfaces, and finally to invade and infiltrate into the adjacent tissue and multiply, the development of secondary tumours being associated with growth of a vascularized host stroma.

Various models have been developed for studies on the pathogenesis of metastasis and for following the effects of chemotherapeutics upon the invasive capacities of neoplastic cells into the normal host tissue. Of interest is the possible extension of such studies to attempts at prognosis based on definition of the invasive and metastatic properties of tumours within a host tissue. Such an approach may have the potential of providing a more selective guide to the clinical progression of tumours, and thereby extend the prediction of the biological behaviour of tumours in vivo now based largely on the morphological and histological features of the neoplastic tissues with malignancy being judged partly on the histologically detectable evidence of invasion.

While animal models have been largely used in investigations of tumour metastasis application of the in vitro approach remains relatively limited despite a long history dating back to the early 1900s [139]. Comprehensive reviews of the application of in vitro systems to the problems of invasion and metastasis have been previously given by *Leighton* [139], *Sigot-Luizard* [224] and *Easty* [52]. Attention has focused on the potential of in vitro models both on the basis that analysis of neoplastic cell behaviour within tissues in vivo may be complicated by vascular reactions and by cellular, neural, hormonal and immunological homeostatic mechanisms, and that the maintenance and upkeep of animals, the possible need for immunosuppression and the occasionally extended periods for tumour formation can be of disadvantage. Within the area of in vitro studies, there has been considerable work on the interaction between neoplastic and normal cells in vitro using cell monolayer culture systems, though some uncertainty has been expressed as to whether such interactions can relate to those in situ. A more appropriate model for the investigation of many of the processes involved in tumour invasiveness could be that of organotypic culture where attempts at the retention of the three-dimensional histoarchitecture of tissues and of different cell and tissue components clearly provide a situation more comparable to that in vivo for the study of, at least, certain aspects of tumour invasion. Furthermore, maintenance of the geometrical configuration of different cell populations in organ culture may be usefully considered in relation to the various discussions on the influence of geometry on differentiation and proliferation of normal and neoplastic cells [73, 75].

Early studies by *Wolff* and co-workers [274, 279] showed that tissue chimaeras were formed when different embryonic organs from the same species, or homologous or heterologous embryonic organs from different species were cultured together. Extension of this approach to heteroplastic associations between neoplastic tissues and embryonic organs led to the tumour-chick embryo mesonephros co-culture technique previously discussed (see Sections II, III, A); while, the vitelline membrane modification provided an elegant demonstration of a transfilter interaction between normal embryonic and neoplastic cells, the nature of which remains to be determined. These studies provided also interesting evidence of the invasive and destructive properties both of human and animal neoplastic tissues and various neoplastic cell lines with a good correlation being observed between the malignancy in host animals and their invasiveness within the co-cultures of chick embryo mesonephros. Direct association of tumour and normal tissue resulted in a rapid fusion of the two tissues, formation of a solid mass of cells and, in general, invasion and replacement of the embryonic tissue by neoplastic cells.

Other studies of similar design have been undertaken to assess and compare the neoplastic potential of a variety of non-neoplastic and neoplastic cells (either primary explants or established cell lines and include those of (a) *Easty and Easty* [53] using new-born or adult mammalian organs (mouse kidney, liver, lung or skin) on the basis that the complication of cell division or morphogenetic changes within the host system would be minimized in post-natal tissues; (b) *DeRidder* et al. [42, 43], *Mareel and De Brabander* [152], *Mareel* et al. [153] and *Pourreau-Schneider* et al. [193] using organ cultures of 7–9 day chick embryo skin and stomach connective tissue, mesonephros, heart, liver, lung and tibiae; (c) *Schleich* [215] and *Schleich* et al. [216, 217] using human decidua graviditatis or human endometrium as the host tissue; and (d) *Noguichi* et al. [178] using 9-day chick embryo skin in organ culture on the basis that such a system may closely mimic the subcutaneous inoculation of cells into animals and that growth of the inoculated cells might be considered equivalent to tumour formation in an animal.

However, the appropriateness of this type of experimental design, which generally involves the immediate confrontation of neoplastic cells with cut edges of normal tissue, has been questionned [49]. In the areas of apposition there is inevitably cell damage and stimulation of tissue regenerative processes which could possibly alter the processes of invasion, although this problem could in part be overcome by preincubation and stabilization of the host tissue [217]. As a consequence, in vitro systems have been designed using as normal target tissue either naturally occurring membranes or tubular structures. The basic consideration was that tumour cell suspensions would be added to the uncut, unwounded surfaces of such target tissues thereby providing a more standardized starting point and facilitating potential quantification [49, 54, 190]. Among the systems explored have been rat or embryonic avian omentum, rat scrotal sac, avian or human amnion [49], chorioallantoic membrane [3, 4, 49, 89, 190] and chick blastoderm [42, 43, 131, 154, 155]. Adult mouse bladder or canine blood vessel [190] and canine thyroid glands, or new-born canine, 7–12 week germ-free rat or human fetal intestine have also served as recipients for tumour biopsy tissue or neoplastic cell lines [70, 74]. In the latter study the tumour explants were implanted either beneath the capsule of the thyroid, injected into the submucosal layer of the intestine or placed on the exposed adhesive mucosal surface of the intestine. In general, the tissue combinations have been assessed on the basis of histological criteria (viz. changes in the organization of the test tissue, pattern of invasion into the target tissue, extent of degenerative changes within the host culture). Quantification of malignant invasion in vitro has been attempted by an adaptation of the $^{51}$Cr-release cytotoxicity test to organotypic cultures [153] while quantitative differences in the depth of invasion between non-neoplastic and neoplastic cell lines have been compared histologically in a limited series of studies [43, 49, 54]. *Noguichi* et al. [178] have attempted, in addition, to use a marker other than invasion by an assessment of the ability of inoculated cells to proliferate in the host tissue system.

As noted by *Wolff* [274] and in several of the above studies [43, 152, 153, 216, 217], tumour invasion in organ culture systems can result in progressive replacement of the host tissue, with the connective tissue often showing a preferential degenerative change in contrast to the relative resistance of the various epithelia. Release of tissue-destructive enzymes (e.g. lysosomal hydrolases and collagenases) has been implicated as one possible mechanism in tumour invasion [189]. Of interest in this context is the study of *Dresden* et al. [48] in which organ culture of a variety of tumours demonstrated a very high frequency of collagenolytic activity by some types of tumours (e.g. tumours of the colon, and carcinomas of both squamous and basal cell origin), while neoplasms of mesenchymal origin, non-neoplastic tissues and other types of epithelial neoplasms only rarely produced collagenase. Assessment of the differential adhesion of neoplastic and non-neoplastic cells to organ cultures of embryonic tissues [42], the depth of invasion [43, 49, 54], the pathway of invasion and the degree of destruction [43, 53, 152, 216, 217, 224] have shown these features to vary considerably according to the biological characteristics of the particular target or test tissue. Although tissue destruction following invasion by neoplastic cells may be clearly demonstrated histologically, quantification of the process by $^{51}$-Cr-release has proven unsuccessful [153]. By contrast, there has been some evidence to suggest that the ability of cells to form localized growths (either benign or neoplastic) could potentially be assessed and quantified in vitro. *Noguichi* et al. [178] showed that while non-neoplastic cells demonstrated an absence of mitotic activity in short-term cultures of chick embryo skin, neoplastic cell lines exhibited extensive and consistent growth in the co-culture system and which correlated with proliferation and tumorigenicity in vivo. Within this context should be included the evidence of gross and microscopic developments of 'microtumours' following co-culture of normal and neoplastic tissues demonstrated by *Folkman* et al. [74] and *Folkman* [70] using canine thyroid or human intestine as target tissues.

The study of neoplastic tissues in matrix culture (see Section II) has shown obvious potentialities in that differences in proliferative capacity, differentiation and growth patterns between normal and neoplastic tissues as well as between individual tumours within pathological groups can, at least in part, be readily identified in such systems [139, 142] while confrontation between heteroplastic associations of normal and neoplastic tissues can impose histopathological changes similar to those described above in other in vitro systems. The extensive studies of *Leighton* [139] and *Leighton* et al. [142] have clearly established the marked capacity of malignant neoplasms to invade matrix lacunae and co-cultured host tissues. Striking differences between the behaviour of benign and malignant epithelial neoplasms have been further stressed with regard to the relatively organized and limited growth of thyroid, parathyroid, renal and colonic adenomas on fibrin foam matrix systems by contrast to the proliferation and invasive tendency of the carcinomas [119, 120]. Similarly, adenocarcinomas of colon and rectum and adenomatous polyps with areas of in situ carcinoma have demonstrated invasive growth whereas no invasive growth may be observed from benign polyps [118]. Further observations, for example on human [207, 208, 226, 230] and experimental [227] tumours involving the central nervous system have confirmed the relationship between invasiveness in vitro and in vivo. Similarly, the potential clinical relevance of the in vitro infiltrative growth characteristics of tumours on fibrin foam matrix cultures has been further suggested from studies of benign intracranial meningiomas [211]. Comparison of the in vitro growth pattern of meningiomas, neurinomas and gliomas showed the meningiomas to be the most active growers and invading most deeply into the lacunae of the matrix. It was suggested that the well-known in vivo infiltrative growth capacity of the meningiomatous cells into surrounding bone may be an inherent characteristic of the neoplasm reproducible under in vitro conditions. Further, in a recent study of human bladder tumours [142] different and distinct patterns of growth and of invasion were established, on the basis of an extensive histological index, from matrix cultures and chick embryo heart co-cultures of such tumours. The study gave evidence of a marked biological heterogeneity of morphological expression in tumours of similar histopathology but the prognostic correlation of these data has yet to be established.

The heterogeneous nature of neoplasms with respect to a variety of biological characteristics has become increasingly documented as discussed earlier [65, 69, 187, 191] including that of metastatic heterogeneity. In vitro models for study of metastatic variant cells have been based, in general, on cell monolayer-type systems though, as discussed by *Leighton* [139] such systems must be considered an oversimplification of the in vivo situation. Of interest, therefore, is a recent organ culture development in which the ability of neoplastic tissues to invade host tissues maintained in organotypic-type systems has been exploited to isolate neoplastic cell lines with increased or decreased metastatic capabilities [190]. The methodology allows recovery of invasive cells (B-16 melanoma lines) following their penetration and passage through various host tissues (chorioallantoic membrane, mouse bladder or canine blood vessel) and isolation within a second, basally-located tissue culture chamber together with a quantification of neoplastic cell invasion. Selection of subpopulations of cells with increasing invasive properties was found to be provided by repeated passage through this organ culture system, the invasive variants selected in the different tissues varying significantly in their properties, despite being derived from the same parent cell line. The various sublines selected in vitro for increased invasiveness in bladder or vein also displayed a greater ability to form spontaneous metastases in vivo than did their parental cells. Chorioallantoic membrane-selected sublines were obtained in which invasiveness and metastatic ability were functionally dissociated whereas variant sublines selected in bladder and vein showed a coupling of these traits.

The interrelationships between neoplastic cells and various cells of the reticuloendothelial system are of importance with respect to the invasive spread of tumours and while widely studied in cell monolayer remain little explored in organ culture. *Nissen and Tanneberger* [176], in a study of the in vitro behaviour of human tumour explants and autologous lymphocytes, determined DNA synthesis in neoplastic tissue and lymphocyte cultures prior and after mixed organ culture. Of the 41 tumours (20 mammary, 17 gastric and 4 bronchial carcinomas) examined, 22 tumour organ cultures were considered to show satisfactory in vitro maintenance on the basis of the DNA synthesis values. Confrontation of the two tissue systems resulted in a stimulation of lymphocyte DNA synthesis in 50% of the mixed organ cultures while 7 of the tumours showed a pronounced reduction of DNA synthesis in the presence of lymphocytes. As discussed by these authors, several technical variables have to be considered in such a system including the lymphocyte ratio to neoplastic cells which is difficult to quantify in the organotypic-type cultures, also the appropriate period of co-culture incubation. Lymph node-breast carcinoma interrelationships studied in vitro by *Gewant* et al. [84] demonstrated a partial or complete inhibition of adenocarcinoma growth in the presence of lymphocytes harvested from axillary lymph nodes, though no inhibition was observed in tumour explants exposed to lymphocytes from nodes involved with metastatic disease. In a more recent study [123] co-culture of stage I and II breast tumours and their axillary lymph nodes were shown to demonstrate a progression of lymphocyte migration from the nodal explants to the tumour explants. This was followed by lymphocytic infiltration in and around the tumour explants, and cytotoxic changes within the neoplastic tissue concomitant with lymphocyte depletion in the lymph node explants. The lymphocytic migration was considered to reflect an immune response towards antigens of the tumour explant. The evidence from these various studies has suggested that co-culture of neoplastic tissue and lymph node explants could be potentially useful in determining immunologically individual tumour-host relationships though this remains to be substantiated.

Attempts to influence the invasive capacity of neoplastic cells by altering either the cell surface properties or the cytoplasmic microtubular function have been attempted in a limited number of studies. The effect of the cationic poly-electrolyte poly-*L*-lysine and the basic dye toluidine blue on invasion of foetal mouse heart by BHK21 Py cells was studied by *Yarnell and Ambrose* [280]. No evidence was found to support the concept that the degree of contact inhibition is governed by the presence of charged molecules at the surface, and that reduction of this charge will prevent invasion in vitro. The invasive properties of the cells were, however, impaired following treatment with low concentrations of poly-*L*-lysine, and also after pretreatment with neuraminidase until the reformation of the surface sialic acid [281].

In another group of studies, *Mareel and De Brabander* [152] found that the microtubule inhibitors colchicine, demecolcine, vincristine sulphate, vinblastine sulphate or methyl[5-(2-thienycarbonyl)-1H-benzimidazol-2-yl]-carbamate totally inhibited neoplastic cell invasion of chick embryo heart organ cultures in co-culture with mouse sarcoma virus transformed cells ($MO_4$). Comparison of the effects of various antimetabolic and antimicrotubular mitostatic drugs on the invasive and proliferative capabilities of the cells demonstrated a differential action. Treatment with antimetabolic mitostatics such as 5-fluorouracil, cytosine arabinoside and bleomycin while leading to an absence of mitotic figures in the co-cultures did not produce an anti-invasive effect in contrast to that imposed by the anti-microtubular compounds and associated with varying mitostatic effects. A functional dissociation of the proliferative and metastatic characteristics of the neoplastic cells was suggested on the basis of these data [152].

### C. Tumour Response to Chemotherapeutic Agents and Ionizing Radiation

In vitro systems have potential scope in screening tumour chemotherapeutic agents, a key advantage of tissue culture being that it can be used to study such agents in human tissues thereby minimizing species differences imposed by whole-animal testing strategies. This, together with the objective, as expressed by *Tanneberger* [239], 'to individualize tumour chemotherapy on the basis of pretherapeutic in vitro sensitivity tests of tumours' has generated a considerable literature. Much of the work has been designed to find out the part of the neoplastic cell cycle at which these agents act, so that useful combinations may be made, either with other drugs or with irradiation [30, 100].

Cell monolayer or primary explant systems have formed the basis of these tissue culture tests, although some attempt has been made to explore the effectiveness of organotypic culture systems for evaluating the response of neoplastic tissues to chemotherapeutic agents [31, 46, 47, 50, 74, 116, 177, 240, 245, 246, 282] or ionizing radiation [40]. That there are basic difficulties in designing appropriate tissue culture tests has been extensively discussed [36, 46, 47, 81, 177]. A central problem of the in vitro tumour chemotherapy prediction approach has been the correlation between the in vitro results and the in vivo effectivity of the particular therapy strategy under study. It has been argued [178] that none of the proposed in vitro prediction tests sufficiently reflects the in vivo situation, namely the biological heterogeneity of cells populating both human and animal neoplasms (see also Introduction), the proliferative and non-proliferative neoplastic cell fractions, the differences in cell kinetics with ageing, and the pharmacokinetics of chemotherapeutic agents. *Dickson and Suzangar* [46] have commented upon the relative advantages of, and factors to be considered in using cell monolayer – or organotypic-type culture systems in chemotherapy prediction assays. In cell monolayer the potential heterogeneity of primary tumour cultures, and selective survival and adaptation to culture of cells from these neoplastic tissues has raised the question as to whether such cells are representative of the neoplastic cell population and whether they represent proliferating or non-proliferating tumour compartments. The development of cultures of defined cell types, both from normal and neoplastic tissues, with retention of differentiated structure and function could provide more precise assay systems for general observations on in vitro responses to cytostatic drugs over a range of cell types. Nevertheless, such systems may be of limited value in attempts to define the specific drug response of neoplastic cell populations within tumours of individual patients. A further consideration is the possible specificity and influence of the stroma on neoplastic cell behaviour and response to chemotherapeutic agents – though this remains to be established. While, the choice of conditions for in vitro assay (such as age and density of the cell culture, also the drug concentration and exposure time employed) can markedly affect the sensitivity of the cell monolayer-type system [81].

Organ culture has the possible advantage of offering a situation whereby tumour chemotherapy and radiation regimes can be explored within an integrated study of the tumour encompassing various intrinsic parameters, namely heterogeneous cell populations, physiological systems and biological processes, operative within neoplastic tissues. Various factors, involving problems of methodology and analysis, may, however, introduce inconsistencies into organ culture assays and need to be more precisely defined to enhance the reproducibility of such assay systems. Among such factors previously discussed in Sections II and III, further reference should be made: (a) to the inherent problem of tissue sampling as a consequence of the heterogeneous nature of neoplastic tumours (which may be controlled in part by prior examination of frozen sections of the biopsy material); (b) the possible selective effect of the in vitro environment on cell and tissue viability, histoarchitecture and growth, and on expression of metabolic function (which may be controlled by further comprehensive studies of media design and morpho-functional life-histories of both normal and tumour organ cultures); (c) the problems of evaluating viability and growth of tumours on the basis of histological surveys or various indices of DNA synthesis, the data being frequently difficult to quantify as a result of the morphological heterogeneity of the tissue sampled and the inhomogeneity in uptake and distribution of radioactive-labelled DNA precursors (which to some extent may be controlled by replicate studies from a large number of explants); and (d) the potential problem of limited or variable diffusion of chemotherapeutic agents into the tumour explants resulting in a variation in effect in different regions of the culture.

In an early attempt to develop an organ culture assay system for the study of the response of human tumours to chemotherapeutic agents, *Yarnell* et al. [282] measured the incorporation by cells of $^{32}$P-labelled phosphate into nucleic acid as an index of growth. A study of myxofibrosarcoma of thigh maintained for 48 h in organ culture indicated a reduced incorporation of labelled phosphate in cultures treated with either methotrexate or melphalan. In a similar study, *Tchao* et al. [245] demonstrated the toxic effects of melphalan and thiotepa on short-term (3- to 4-day) organ cultures of mouse sarcoma 180, the effect of thiotepa becoming operative only some 48 h after treatment of the cultures. In both series of studies, the chemotherapeutic agents were usually added a few hours after the cultures were set up. The inhibitory effects of mannitol-myleran, melphalan, thiotepa and methotrexate were also assessed on a series of human tumours (melanoma; carcinoma of the ovary and maxilla; glioma; sarcoma) both by measuring the level of $^{3}$H-thymidine incorporation into the DNA and by histological criteria which, in general, showed good agreement [246]. None of the chemotherapeutic agents showed any specificity of action on the tumours, also the tumours varied considerably in their sensitivity to any one drug. However, it was found that the order of sensitivity of any tumour to the series of test compounds was similar and led to the suggestion that this type of in vitro assay could be of use in assessing the relative sensitivity of individual tumours to a range of chemotherapeutic agents.

*Dickson and Suzangar* [46] discussed a short-term organotypic-type culture system in which tumour sensitivity to a chemotherapeutic agent was defined as a minimum 30% inhibition of the major energy-yielding processes of the neoplastic cell (respiration and anaerobic glycolysis), associated with a minimum 50% reduction of isotope precursor uptake into DNA, RNA or protein. Such inhibition of metabolism, it was suggested, could be interpreted to imply the presence of a drug-sensitive population of neoplastic cells, and furthermore, by using a multi-parameter assessment for the assay system, the predictive potential of such systems could be enhanced. This approach was explored in a rat (Yoshida sarcoma) model system, in which an attempt was made to correlate the minimal drug dose that inhibited tumour growth in vivo with the degree of metabolic inhibition produced in culture by a calculated equivalent in vitro dosage, and this assay system was extended to a series of human tumours. It was concluded from these studies that, while a considerable inhibition of the different parameters measured in the neoplastic tissues could be obtained in response to various chemotherapeutic agents, such data was to be interpreted with caution. Mitigating against an extrapolation of the results to the in vivo situation could be the heterogeneous nature of tumours and the inherent problems of tumour sampling and of tumour mass behaviour. Furthermore, assay of the human neoplastic tissues showed no correlation between the histological classification of any of the human tumour types examined and the rate of metabolism or response to specific cytotoxic drugs, while large variations in the response to drugs of individual tumours of all types was observed as previously discussed.

In a further study to assess the clinical value of organ culture drug prediction assays [177], two randomized trials of lung and ovarian carcinomas comparing predicted, non-predicted and no chemotherapy treatments (trenimon, vinblastine, daumoblastin, 5-fluorouracil and methotrexate) were carried out. DNA synthesis values, determined from liquid scintillation counts, were used to assess the effect of these drugs on short-term (48 h) organ cultures of the neoplastic tissues. No statistical significant difference between the organ culture predicted and non-predicted regimes with respect to patient survival rates could be demonstrated and this was accounted for by the inherent problems of the assay system as discussed earlier and by the fact the drug responses of normal and neoplastic cell populations were not distinguished.

Short-term prostate organ cultures from normal rat or pathological (BPH and carcinoma) tissues have been utilized in the development of a model in vitro assay system for testing chemotherapeutic agents against prostatic neoplasms [116]. A biochemical parameter was used with changes in 5α-reductase enzyme activity (5α-RA) in homogenates of the prostatic tissue explants being found to provide a sensitive and direct assay of the effects of various agents on the in vitro target tissue. It was shown that estracyst, 17β-estradiol, progesterone and other steroid conjugates could have markedly inhibitory effects on 5α-RA in the short-term culture of rat prostate. However, in human BPH or prostatic carcinomas a wide range of effects was produced and it was suggested that these differences in responsiveness to steroidal agents may be accounted for by differences in the type of prostatic lesion and degree of sensitivity to hormonal action. It was concluded that while 5a-RA as the sole index of hormone or drug action may be of limited value, it could be usefully incorporated into a multiparameter assessment of the effect of such agents on prostatic neoplasms.

Little attempt has been made to explore the use of organ culture systems for establishing the response of neoplastic tissues to ionizing radiation. *Dendy* [40] and *Beaupain* [14–16] have, however, demonstrated that colon carcinoma cultures, when subjected to a single or double dose of radiation of between 100 and 800 rad could show progressive changes in mitotic index and increasing necrosis and neoplastic cell death.

From the evidence currently available, the potential of organotypic-type systems for the study of neoplastic tissue response to chemotherapeutic agents or to ionizing radiation would appear limited. Whether a more encouraging future may be offered to this type of experimental system with further characterization of the methodological and analytical parameters of tumour organ culture has yet to be determined. Nevertheless, there is evidence [140] that some attention should be directed to such model systems in which an attempt is made to consider the tumour in its totality and in which certain modalities of drug and radiation effects may be more appropriately explored.

### D. Tumour Response to Virus Infection

It is possible by the use of organ culture to reproduce in vitro virus disease processes comparable to those observed in vivo. This was initially demonstrated in investigations on the response of embryonic trachea and skin to influenza, herpes simplex and vaccinia viruses [13], and of chick embryo skin to fowl pox virus [276, 277]. Subsequent studies on the response of mouse salivary gland and metanephric rudiments to polyoma and vesicular stomatis viruses [34, 213]; of skin to herpes simplex and vaccinia viruses [101, 102, 163, 199], and of trachea to myxoviruses [109], clearly established the wide application of organ culture to various problems of host-virus interaction.

A more substantial literature has since evolved and for further discussion, reference should be made to previous reviews [35, 110, 196]. From the various investigations have been demonstrated the use of organ culture systems for the isolation and cultivation in vitro of viruses where cell monolayer cultures have proven refractory; for the assessment of virulence, attenuation or selection of viruses; for examining the influence of cellular differentiation and of morphogenesis on tissue susceptibility to virus infection; for the study of the pathological effects of viruses on tissues and of the ability of organ cultures to respond to virus infection by producing interferon; and for assessing the effects of anti-viral drugs and of interferon on virus-infected organ culture models.

While there has been little work on oncogenic viruses in organ culture, a recent study has compared the susceptibility of normal, dysplastic and neoplastic (carcinoma in situ and invasive carcinoma) human cervices to herpes simplex type 2 virus (HSV-2), and also the induced cytopathological alterations [151]. As discussed by these authors, a complex relationship between HSV-2 and premalignant and malignant lesions of the uterine cervix has been suggested. The study showed that following HSV-2 infection of organ cultures from normal ecto- and endocervical tissue, changes characteristic of herpetic infection occurred and which were very similar to those described in herpetic cervicites. Dysplastic explants when infected with HSV-2 presented also essentially the same cytopathological alterations, but CIS and invasive carcinoma explants, while responding in an identical way, in addition showed a large number of polykaryocytes. There is evidence that this type of organ culture model might give help in the study of potential factors contributing to the malignant progression of pathological lesions of the cervix, as has been also suggested from a similar study by *Tobin* et al. [248].

In an earlier approach to the possible response of pathological tissues to viruses, organ cultures of various human gastric carcinomas (stomach and rectum) were tested as a model for the study of replication of different strains of poliovirus presumed to possess oncolytic properties [255]. The data demonstrated the ability of poliovirus to multiply in these cultures and indicated some degree of oncolysis. In a subsequent and comparative study of poliovirus-infected and control gastrointestinal tumour organ cultures it was demonstrated that wild poliovirus strains exhibited the greatest oncolytic activity, whereas, of the attenuated (vaccine) Sabin strains, type III produced a distinct though less marked effect, while type II showed no oncolytic properties [251]. The oncolytic potentials of viruses for human cancer have been only intermittently explored but, as suggested from these studies, it is possible that organ cultures, in retaining neoplastic tissue characteristics, could provide a useful system for the study of virus oncolysis in situations where such an approach may be considered appropriate.

## V. Concluding Comments

Tumours, as found in their solid state, present an organized character involving the coordinated participation of various tissue components (neoplastic cells, interstitial tissue and vascular system). As a consequence, attention may be directed to a consideration of the tumour in its totality not only from the standpoint of the biology of tumour growth and function, but also with regard to therapy. Emphasis on in vitro model systems that seem applicable to the retention of the organized character of tumours has led to the application of organ culture methodology. In providing a system in several ways closer to the in vivo situation, but allowing for better control of the environment than in the animal, organ culture is of considerable interest with regard to controlled investigations of human tumours which cannot be carried out in vivo.

The possibility of the maintenance of tumours in organ culture with at least some retention of their original structures, levels of differentiation and functional properties has been illustrated in a number of studies as discussed in this review. As a consequence, it has been suggested that such an in vitro system could be particularly suited for the study of the histological organization of tumours, allow valid correlations between in vivo and in vitro morphologic expressions of differentiation and anaplasia and serve as a prognostic tool. The organ culture of tumours has also been assumed to provide information about the proliferation kinetics of in vivo neoplastic cells, to facilitate the definition of their invasive (and metastatic) properties, and to exemplify behavioural response to hormones or chemotherapeutic agents from which useful therapeutic suggestions could be derived.

The proficiency of organ culture for investigative studies of tumour behaviour may, however, be considered still debatable in that adequate in vitro maintenance and representation of the major components of neoplastic tissues is in general limited. Inevitably, this poses the question as to whether the behavioural characteristics of such tissues in vitro merely reflect the variability of the method or whether they reflect accurately the behaviour of tumours in vivo. Long-term survival of tumours in organ culture and the retention of many aspects of growth, differentiation and function would make possible meaningful studies following exposure to hormones or chemotherapeutic agents, of value in providing a baseline for future therapeutic approaches. Within the intrinsic limitations of the system a fuller understanding of the variables of tumour organ culture system with regards to methodology and analysis must be imposed if in vitro organ culture studies of tumours are to be relevant. While organ culture may serve as a simple model for the study of biological properties of tumours, major advances in its future use in this area are only likely on the basis of fresh development of techniques both as regards maintenance of the tissue itself, and the methods by which tumour organ cultures may be analyzed.

*Appendix.* A survey of tumor organ culture studies

| | | | | | |
|---|---|---|---|---|---|
| A | Plasma clot | BSS | Balanced salt solution | L | Lymphocyte |
| B | Solid agar | CA | Chemotherapeutic assays | LE | Liver extract |
| C | Floating raft | CEE | Chick embryo extract | LED | Dialysate of liver extract |
| D | Platform | CMC | Carboxymethylcellulose | K | Karyotype analysis |
| E | Perfusion | CO | Corticosterone | MCA | 3-Methylcholanthrene |
| F | Submerged | CP | Chicken plasma | NS | Not stated |
| G | Sponge foam matrix | DE | Dextrose | S | Serum |
| H | Fibrin foam matrix | DMBA | 7,12-dimethylbenz(a)anthracene | TA | Tumour plasminogen activator assay |
| AA | Ascorbic acid | HC | Hydrocortisone | TH | Test hormones |
| AL | Albumin | GL | Glucose | TP | Tripeptide supplement |
| ALN | Axillary lymph nodes | GP | Glycoprotein synthesis | TT | Tumorigenicity test |
| AN | Immunologic analysis | I | Insulin | V | Virus infection |
| BEE | Beef embryo extract | IND | Inducer tissue | X | X-radiation |

| Tissue/Tumour | Species | Organ culture system | Medium | Gas phase %CO$_2$ | %O$_2$ | Culture period | Reference(s) |
|---|---|---|---|---|---|---|---|
| *Cervix* | | | | | | | |
| Carcinoma | human | D + Millipore support | Eagle's + S (10%) | 10 | 90 | 4 days | 201 |
| Squamous cell carcinoma | mouse | G | Eagle's + S (30%) | NS | | 14 days | 144 |
| Carcinoma | mouse | D | MEM [+ GP: AN] | 5 | 95 | 30 h | 25 |
| Carcinoma | human | B | Eagle's BSS + S (20%) + CEE [+ TT] | NS | | 2–8 days | 178 |
| Squamous carcinoma | human | D | MEM + S (10%) + CMC (1.5%) [+ V] | 5 | | 1–8 weeks | 151 |

| Tissue/Tumour | Species | Organ culture system | Medium | Gas phase %CO$_2$ | %O$_2$ | Culture period | Reference(s) |
|---|---|---|---|---|---|---|---|
| *Endocrine system* | | | | | | | |
| Thyroid: adenoma | human | A:D | CP + CEE: | 0.4 | 20 | 7–19 days | 204 |
| | | | or CMRL 1066 | 5.0 | 1 | | |
| | | | or Eagle's s + S (20%) | 5.0 | 95 | | |
| Thyroid: adenocarcinoma, papillary carcinoma | human | D + Millipore support | Eagle's + S (10%) | 10 | 90 | 11 days | 201 |
| Thyroid: adenocarcinoma, papillary carcinoma | human | C: D + cellulose acetate support | 199 + S (10%) | 5 | | 6 days | 5 |
| Thyroid: adenoma, carcinoma | human | F | 199 | 5 | 95 | 20 days | 175 |
| Pituitary: adenoma | human | D + lens paper support | 199 + S (20%) | | | 7–28 days | 184, 185 |
| Thyroid: papillary carcinoma | human | B | Eagle's + S (10%) + CEE + YE ± mesonephros | NS | | 6–12 days | 160 |
| Parathyroid: adenoma | human | D + lens paper support | MEM + S (10%) | 5 | 95 | | 60 |
| *Endometrium* | | | | | | | |
| Adenocarcinoma | human | G | 199 + S (10%) [+ TH] | NS | | 8 days | 179 |
| Rhabdomyosarcoma | human | G | | | | | |

| Tumour | Species | Type | Medium | | | Duration | Ref |
|---|---|---|---|---|---|---|---|
| Rhabdomyosarcoma | human | B | 199 + S (10%) [+ TH] | 5 | | 3–4 days | 126, 127 |
| Rhabdomyosarcoma | human | B | Eagle's + S (25%) + GL + YE [+ TH] | NS | 95 | 9 days | 158 |
| Carcinoma | human | G | 199 + S (20%) + CEE [+ TH] | NS | | 9 days | 180 |
| Carcinoma | human | D + lens paper support | 199 + S (30%) + I [+ TH] | 5 | | 4 days | 114 |
| Carcinoma | human | H | 199 + S (20%) + CEE | 5 | | 7 days | 229 |
| Carcinoma | human | D + lens paper support | 199 + S (10%) + GL + I [+ TH] | 5 | 95 | 7 days | 111 |

*Gastrointestinal system*

| | | | | | | | |
|---|---|---|---|---|---|---|---|
| Large intestine adenocarcinoma, rectum adenocarcinoma | human | A:D | CP + CEE; or CMRL 1066; or Eagle's + S (20%) | 0.4<br>5.0<br>5.0 | 20<br>1<br>95 | 7–19 days | 204 |
| Colon adenocarcinoma, rectum carcinoma, rectum adenocarcinoma, stomach adenocarcinoma | human | D + Millipore support | Eagle's + S (10%) | 10 | 90 | 11 days | 201 |
| Stomach adenocarcinoma, colon adenocarcinoma, rectum adenocarcinoma | human | B | Eagle's + S (10%) + CEE + YE ± mesonephros | NS | | 6–12 days | 160 |

| Tissue/Tumour | Species | Organ culture system | Medium | Gas phase %CO$_2$ | %O$_2$ | Culture period | Reference(s) |
|---|---|---|---|---|---|---|---|
| Colon adenocarcinoma | human | G | Eagle's + Hank's BSS + S (30%) | NS | | 7 days | 138 |
| Colon carcinoma | human | G | Eagle's + Hank's BSS + S (30%) | NS | | 14 days | 144 |
| Liver metastasis from gastric primary carcinoma (Z 200) | human | B | Gey's + Tyrode's BSS + CEE + S (20%) + mesonephros (or YE; LE) [+ AN:K:X] | NS | | continuous culture since 1962 | 274, 279 |
| Colon carcinoma Z 516 | human | B | Gey's + Tyrode's BSS + CEE + S (20%) + mesonephros (or YE; LE) [+ AN] | NS | | 10 years | 274, 279 |
| AZ110 | human | B | Gey's + Tyrode's BSS + CEE + S (20%) + mesonephros (or YE; LE) [+ AN:K: X] | NS | | 10 years | 274, 279 |

| | | | | | | |
|---|---|---|---|---|---|---|
| Liver metastasis from carcinoma of rectum (AZ 432) | human | B | Gey's + Tyrode's BSS + CEE + S (20%) + mesonephros (or YE; LE) [+ AN: K: X] | NS | < 5 years | 274, 279 |
| Colon, rectum | human | H | Eagle's + S (10%) | 5 | | 3–21 days | 118 |
| Stomach carcinoma | human | D + Millipore support | 199 + S (20%) [+ CA] | 5 | 95 | 2 days | 240 |
| Stomach adenocarcinoma; rectum: medullary carcinoma, scirrhous carcinoma | human | D + Millipore support | Eagle's + S (10%) ± CEE [+ V] | 5 | | 1–18 days | 250, 251, 255 |
| Oesophagus squamous cell carcinoma | human | D + Millipore support | | | | | |
| Colon carcinoma, rectum carcinoma | human | D | F10 + S (20%) [+ AN] | 5 | 95 | 5–14 days | 23 |
| Stomach carcinoma | human | D + Millipore support | 199 + S (20%) ± lymphocytes [+ AN] | 5 | 95 | 2–4 days | 176 |
| Colon adenocarcinoma | human | B | Eagle's BSS + S (20%) + CEE [+ TT] | NS | | 2–8 days | 178 |
| *Integumentary system* | | | | | | | |
| Melanoma | human | A:D | CP + CEE: or CMRL 1066: or Eagle's + S (20%) | 0.4 / 5.0 / 5.0 | 20 / 1 / 95 | 7–19 days | 204 |
| Melanoma | mouse | D + rayon support | Eagle's + S (10%) [+ TT] | 5 | 95 | 7 days | 53 |

| Tissue/Tumour | Species | Organ culture system | Medium | Gas phase %CO$_2$ | Gas phase %O$_2$ | Culture period | Reference(s) |
|---|---|---|---|---|---|---|---|
| Melanoma | mouse | E | hemoglobin + S [+ TT] | | | 5 days | 74 |
| Melanoma | mouse | B | Eagle's + S (10%) + CEE + YE ± mesonephros | NS | | 6–12 days | 160 |
| Squamous cell carcinoma | mouse | B | Eagle's + S (10%) + CEE + YE ± mesonephros | NS | | 6–12 days | 160 |
| Melanoma | human | D + Millipore support | Eagle's + S (10%) | 10 | 90 | 11 days | 201 |
| Melanoma | human | C: D + cellulose acetate support | 199 + S (10%) | 5 | | 6 days | 5 |
| Melanoma | human | B + Millipore support | TC 109 [+ CA] | 5 | 95 | 3–5 days | 246 |
| *Mammary* | | | | | | | |
| Adenofibroma, adenocarcinoma, scirrhous carcinoma | human | A: D | CP + CEE: *or* CMRL 1066: *or* Eagle's + S (20%) | 0.4<br>5<br>5 | 20<br>1<br>95 | 7–19 days | 204 |
| Carcinoma | mouse | C + lens paper support | 199 [+ TH] | 5 | 95 | 5–10 days | 57, 200 |
| Adenocarcinoma, infiltrating ductal carcinoma, scirrhous carcinoma | mouse | D + Millipore support | Eagle's + S | 10 | 90 | 11 days | 201 |

| Tumour | Species | Method | Medium | | | Duration | Reference |
|---|---|---|---|---|---|---|---|
| Fibroadenoma, carcinoma | mouse | C: D + cellulose acetate support | 199 + S (10%) | 5 | | 6 days | 5 |
| Adenocarcinoma, scirrhous carcinoma | human | B | Eagle's + S (10%) + CEE + YE ± mesonephros | NS | | 6–12 days | 160 |
| Adenocarcinoma | mouse | G | Eagle's + Hank's BSS + S (30%) | NS | | 14 days | 144 |
| Carcinoma | mouse | B | 199 + S (25%) [+ IND] | NS | | 3–10 days | 147 |
| DMBA-induced carcinoma | rat | C | 199 + GL ± I | 5 | 95 | 4 days | 95, 96, 98 |
| DMBA-induced carcinoma | rat | D + Millipore support | 199 + GL [+ TH] | 5 | 95 | 4 days | 99, 181 |
| Carcinoma | human | F | 199 + S (10%) + GL [+ TH] | 5 | | 2 days | 235 |
| MCA-induced carcinoma | rat | D + rayon acetate support | 199 + S (1–10%) [+ TH] | 5 | 95 | 2 days | 237, 238 |
| Adenocarcinoma | mouse | B | BSS + S (20%) + CEE + mesonephros | NS | | 50 days | 274 |
| Carcinoma | rat | B | | | | 30 days | |
| Carcinoma | mouse | D + Millipore support | Eagle's + S (10%) | 5 | 95 | 10 days | 197 |
| Dysplasias fibroadenoma, intraductal papilloma, colloid carcinoma, scirrhous carcinoma | human | D + cellophane screen | 199 + GL [+ TH] | 5 | 95 | 11 days | 261 |

| Tissue/Tumour | Species | Organ culture system | Medium | Gas phase %CO$_2$ | Gas phase %O$_2$ | Culture period | Reference(s) |
|---|---|---|---|---|---|---|---|
| DMBA-induced carcinoma | rat | D | 199 + I + CO [+ TH] | 5 | 95 | 4 days | 267 |
| Fibroadenoma, medullary and intraductal carcinomas, scirrhous carcinoma | human | D + agar support | Eagle's + S (10%) + I [+ TH] | 5 | 95 | 6 days | 270 |
| Dysplasias, fibroadenomas | human | D | 199 ± I | NS | | 20 days | 56 |
| Carcinoma | human | D + Millipore support | 199 + S (20%) [+ CA] | 5 | 95 | 2 days | 240 |
| Adenocarcinoma | mouse | D + Millipore support | Eagle's (+ Earle's BSS) + S (10%) + CEE [+ IND] | 5 | | 6–13 days | 37, 38 |
| Carcinoma | human | D | 199 [+ TH] | 5 | | 1–3 days | 130 |
| Carcinoma | human | D + Millipore support | Eagle's + S (20%) | NS | | 6–10 days | 159 |
| Intraductal carcinoma, scirrhous carcinoma | human | C + Millipore on cellulose acetate support | T8 + AL (1%): or T8 + S (20%): or 199 + S (20%): [+ TH] | NS | | 1–3 days | 8, 9 |
| DMBA-induced carcinoma | rat | C + Millipore on cellulose acetate support | T8 + AL (1%): or T8 + S (20%): or 199 + S (20%): [+ TH] | NS | | 1–3 days | 7 |

# Tumour Growth, Differentiation, Function in Organ Culture

| Tissue | Species | Type | Medium | | | Duration | Ref |
|---|---|---|---|---|---|---|---|
| Fibroadenoma carcinoma | human | D + lens paper support | 199 + S (20%) [+ TH] | 5 | 95 | 2–10 days | 67, 82, 83 |
| Carcinomas, various histological types, including scirrhous | human | D + Millipore or agar support | 199 + GL [+ TH] | 5 | 95 | 14 days | 99, 182 |
| Carcinoma | human | D | Eagle's + S (20%) + BEE [+ GP] | NS | | 2–3 days | 44 |
| Carcinoma | human | F | 199 + I + CO [+ TH] | 5 | 95 | 3 days | 124 |
| Infiltrating duct, scirrhous carcinomas, medullary carcinoma | human | D | T8 [+ TH] | 5 | 95 | 1–3 days | 19, 20 |
| Carcinoma | human | D + Millipore support | 199 + S (20%) + lymphocytes [+ AN] | 5 | 95 | 2–4 days | 176 |
| Carcinoma (spontaneous) DMBA-induced carcinoma | mouse rat | D | 199 [+ TH] | 5 | 95 | 2 days | 263 |
| Carcinoma | human | F | 199 [+ TH] | 5 | 95 | 2 days | 262, 264–266 |
| Benign hyperplasia | human | D + Millipore support | McCoy 5a + S (20%) [+ IND] | 5 | | 10 days | 6 |
| Carcinoma (stage I and II) | human | B | BSS + S + EE [+ ALN] | NS | | 6–8 days | 123 |
| Infiltrating ductal and lobular carcinomas | human | D | 199 + HEPES [+ TH] | 5 | | 10 days | 273 |

| Tissue/Tumour | Species | Organ culture system | Medium | Gas phase %CO$_2$ | %O$_2$ | Culture period | Reference(s) |
|---|---|---|---|---|---|---|---|
| *Miscellaneous* | | | | | | | |
| Androgen-estrogen-induced, dependent flank organ carcinoma | hamster | A: D + lens paper support | CP + CEE + S; *or* CMRL 1066 + S (10–50%) *or* 199 + CEE [+ TH] | 5 | | 4 weeks | 1, 2 |
| Ameloblastoma | human | A:D | CP + CEE: *or* CMRL 1066: *or* Eagle's + S (20%) | 0.4 5.0 | 20 1 | 7–19 days | 204 |
| Fibrosarcoma | human | | | | | | |
| Sarcoma 180 | mouse | D + rayon support | Eagle's + S (10%) [+ TT] | 5.0 | 95 | 7 days | 53 |
| Myeloma | mouse | E | hemoglobin + S [+ TT] | 5 | 95 | 5 days | 74 |
| Mast cell tumour | | | | | | | |
| Myxofibrosarcoma | human | C | Eagle's + S (10%) [+ CA] | 5 | | 48 h | 282 |
| Fibrosarcoma | human | B | Eagle's + S (10%) + CEE + YE ± mesonephros | NS | | 6–12 days | 160 |
| Ascites hepatoma | mouse, rat | G | Eagle's + S (20%) | NS | | 17 days | 138, 144 |
| Sarcoma 180 | mouse | B | 199 + S (25%) [+ IND] | NS | | 3–10 days | 147 |
| Giant cell bone tumour | mouse | | | | | | |
| Giant cell osteogenic tumour | human | | | | | | |

| | | | | | |
|---|---|---|---|---|---|
| Sarcoma 180 | mouse | B + Millipore support | NCTC 109 + S (16–20%) + CEE [+ IND] | 5 | 6–12 weeks 58 |
| Sarcoma S 180 | mouse | B | BSS + CEE + S (20%) + mesonephros | NS | 157 days 274 |
| Myelosarcoma T58 | rat | B | BSS + CEE + S (20%) + mesonephros | NS | 30 days |
| Zajdela's hepatoma | rat | B | BSS + CEE + S (20%) + mesonephros | NS | 7 months |
| Cell lines: | | | | | |
| NCTC 2472 fibroblast | mouse | B | BSS + CEE + S (20%) + mesonephros | NS | 263 days |
| PTT 12 fibroblast | mouse | B | BSS + CEE + S (20%) + mesonephros | NS | 263 days |
| KB epithelial | human | B | BSS + CEE + S (20%) + mesonephros | NS | |
| Hela epithelial | human | B | BSS + CEE + S (20%) + mesonephros | NS | |
| HEp2 epithelial | human | B | BSS + CEE + S (20%) + mesonephros | NS | up to 300 days |

| Tissue/Tumour | Species | Organ culture system | Medium | Gas phase %CO$_2$ | Gas phase %O$_2$ | Culture period | Reference(s) |
|---|---|---|---|---|---|---|---|
| Osgood J111 epithelial | human | B | BSS + CEE + S (20%) + mesonephros | NS | | | |
| Detroit 116P epithelial | human | B | BSS + CEE + S (20%) + mesonephros | NS | | | |
| Detroit 6 epithelial | human | B | BSS + CEE + S (20%) + mesonephros | NS | | | |
| Sarcoma | human | D + Millipore support | 199 + S (20%) [+ CA] | 5 | 95 | | 240 |
| 7,12-DMBA-induced subcutaneous sarcoma | rat | C + Millipore on cellulose acetate support | T8 + AL (1%): or T8 + S (20%): or 199 + S (20%) [+ TH] | NS | | 1–3 days | 7 |
| Virus-induced sarcoma | rat | | | | | 1–3 days | 7 |
| Neoplastic cell line: MO4 | mouse | B | AL + S + CEE + mesonephros or chick embryo heart [+ CA] | NS | | 4 days | 152 |
| Neoplastic cell lines: W, Dr, HeLa, IMR-32, HT-1080 | human | B | Earle's BSS + S (20%) + CEE [+ TT] | NS | | 2–8 days | 178 |

| | | | | | | |
|---|---|---|---|---|---|---|
| Fibrosarcoma | human | B | Earle's BSS + S (20%) + CEE [+ TT] | NS | 2–8 days | 178 |
| Testicular tumours- seminoma, sertolinoma | dog | D | 199 [+ IND] | 5 | 3 days | 186 |
| *Neuronal* | | | | | | |
| Glioma | human | D + lens paper or gelatin foam support | Eagle's + S (10%) | 5 | 4 days | 188 |
| Glioma | human | B + Millipore support | TC 109 [+ CA] | 5    95 | 3–5 days | 246 |
| Meningioma | human | H | Eagle's + S (10%) | 5 | 14 days | 211 |
| Glioblastoma | human | D + G or Millipore support | Simm's BSS + S (40%) + D + I | 3–4 | 122 days (+ Millipore); 137 days (D + G) | 207, 226 |
| Astrocytoma | human | D + G or Millipore support | Simm's BSS + S (40%) + D + I | 3–4 | 29 days (+ Millipore); 43 days (D + G) | 226 |
| Ethylnitrosurea-induced astrocytoma | rat | D + G or Millipore support | Simm's BSS + S (40%) + D + I | 3–4 | 19 days (+ Millipore); 55 days (D + G) | 227 |

| Tissue/Tumour | Species | Organ culture system | Medium | Gas phase %CO$_2$ | Gas phase %O$_2$ | Culture period | Reference(s) |
|---|---|---|---|---|---|---|---|
| Ethylnitrosurea-induced schwannoma | rat | D + G or Millipore support | Simm's BSS + S (40%) + D + I | 3–4 | | 43 days (+ Millipore): 82 days (D + G) | 205 |
| Schwannoma | human | D + G or Millipore support | Simm's BSS + S (40%) + D + I | 3–4 | | 55 days | 205 |
| Neuroblastoma | human | B + mosaic or double membrane | BSS + CEE + S (20%) + LD ± mesonephros | NS | | 13–46 weeks | 148 |
| Ependymoma | human | B + mosaic or double membrane | BSS + CEE + S (20%) + LD ± mesonephros | NS | | 3 weeks | 150 |
| Medulloblastoma | human | B + mosaic or double membrane | BSS + CEE + S (20%) + LD ± mesonephros | NS | | 7 weeks | 150 |
| Schwannoma | human | B + mosaic or double membrane | BSS + CEE + S (20%) + LD ± mesonephros | NS | | 22 weeks | 150 |
| Neuroblastoma | human | B | Earle's BSS + S (20%) + CEE [+ TT] | NS | | 2–8 days | 178 |

*Ovary*

| | | | | | | | |
|---|---|---|---|---|---|---|---|
| Granulosa cell carcinoma | human | D + Millipore support | Eagle's + S | 10 | 90 | 11 days | 201 |
| Papillary carcinoma | human | D + Millipore support | | | | | |
| Carcinoma | human | B + Millipore support | TC 109 [+ CA] | 5 | 95 | 3–5 days | 246 |
| Carcinoma | human | D + Millipore support | 199 + S (20%) [+ CA] | 5 | 95 | 2 days | 177, 240 |
| Adenocarcinoma | human | G | 199 + TP [+ TA] | 5 | | 30 days | 10 |

*Prostate*

| | | | | | | | |
|---|---|---|---|---|---|---|---|
| Benign hyperplasia | human | D + agar support | T8 | 5 | 95 | 9 days | 218 |
| Adenocarcinoma | human | D + agar support | Eagle's + S (10%) + I + AA [+ TH] | 5 | 95 | 4 days | 165 |
| Adenocarcinoma | human | D + agar or cellulose acetate support | Eagle's + S (10%) + DE + I + AA [+ TH] | 5 | 95 | 6 days | 166, 167 |
| Adenocarcinoma | human | D + agar or cellulose acetate support | T 8 | 5 | 95 | 12 days | 87, 88 |

| Tissue/Tumour | Species | Organ culture system | Medium | Gas phase %CO$_2$ | Gas phase %O$_2$ | Culture period | Reference(s) |
|---|---|---|---|---|---|---|---|
| Adenomata | human | F | Eagle's + S (10%) + I | 5 | 95 | 8 days | 168 |
| Benign hyperplasia | human | D + lens paper support | 199 + S (10%) [+ TH] | 5 | 95 | 6 days | 136 |
| Benign hyperplasia | human | D + Millipore support | Eagle's + S (10%) [+ TH] | 5 | 95 | 4 days | 198 |
| Benign hyperplasia, carcinoma | human | D + lens paper support | 199 [+ TH] | 5 | | 2–4 days | 116 |
| Benign hyperplasia, carcinoma | human | F | Eagle's + S (20%) | 5 | | 8 days | 128 |
| *Respiratory tract* | | | | | | | |
| Lung adeno-carcinoma, squamous cell carcinoma | human | D + Millipore support | Eagle's + S (10%) | 10 | 90 | 11 days | 201 |
| Lung carcinoma | human | C: D + cellulose acetate support | 199 + S (10%) | 5 | | 6 days | 5 |
| Lung adeno-carcinoma | human | B | BSS + S (20%) + CEE + mesonephros | NS | | 15.5 months | 274 |
| Lung carcinoma | human | D + Millipore support | 199 + S (20%) [+ CA] | 5 | 95 | 2 days | 177, 240 |
| Bronchial carcinoma | human | D + Millipore support | 199 + S (20%) ± lymphocytes [+ AN] | 5 | 95 | 2–4 days | 176 |

## Salivary gland

| | | | | | | | |
|---|---|---|---|---|---|---|---|
| Parotid: squamous cell carcinoma; submaxillary: squamous cell carcinoma | human | D + Millipore support | Eagle's + S | 10 | 90 | 11 days | 201 |
| Parotid: mixed tumour | human | B | Eagle's + S (10%) + CEE + mesonephros | NS | | 6–12 days | 160 |
| | | + YE ± | | | | | |
| Carcinoma | human | C: D + cellulose acetate support | 199 + S (10%) | 5 | | 6 days | 5 |
| Parotid: mixed tumour | human | B | 199 + S (25%) [+ IND] | NS | | 3–10 days | 147 |
| Parotid: carcinoma | human | A:D | CP + CEE: or CMRL 1066: or Eagle's + S (20%) | 0.4<br>5.0 | 20<br>1 | 7–19 days | 204 |
| Teratocarcinoma | mouse | B + double membrane | BSS + CEE + mesonephros: or Ham's F12 + mesonephros | 5.0<br>NS | 95 | NS | 59 |
| Teratocarcinoma | mouse | D + G<br>D + Millipore | Simm's BSS + S (40%) + D + I: or Simm's BSS + Eagle's + CEE + DE + I | 3–4 | | 48 days (Millipore)<br>68 days (D + G) | 252 |
| Teratocarcinoma | mouse | D + Millipore support | Eagle's + S (15%) | 5 | | 30 days | 33 |

## Urinary system

| Tissue/Tumour | Species | Organ culture system | Medium | Gas phase %CO$_2$ | %O$_2$ | Culture period | Reference(s) |
|---|---|---|---|---|---|---|---|
| Kidney estrogen-induced dependent tumour | hamster | A: D + lens paper support | CP + CEE + S or CMRL 1066 ± S (10–50%) [+ TH] | 5 | | 4 weeks | 1 |
| Ureter transitional cell carcinoma | human | D + Millipore support | Eagle's + S | 10 | 90 | 11 days | 201 |
| Kidney carcinoma | human | C: D + cellulose acetate support | 199 + S (10%) | 5 | | 6 days | 5 |
| Kidney carcinoma | rat | B + Millipore support | NCTC 109 + S (16–20%) + CEE [+ IND] | 5 | | 6–12 weeks | 58 |
| Kidney nephroblastoma | human | D | 199 + S (10%) + CEE [+ IND] | 5 | 95 | 2 months | 32 |
| Kidney nephroblastoma | human | D + lens paper or Millipore support | Eagle's + S (10%) | 5 | | 3–7 weeks | 202 |
| Kidney nephroblastoma | human | D + lens paper or Millipore support | Eagle's + S (10%) ± mesonephros [+ IND] | 5 | | 3 weeks | 203 |
| Bladder carcinoma | human | G | Waymouth's MB 752/1 + S (20%) | NS | | 1 week | 142 |

| | | | | | |
|---|---|---|---|---|---|
| Bladder carcinoma | human | D + Millipore support | Waymouth's MB 752/1 + S (10%) + AA + HC | 5 | 4 weeks | 105 |
| Bladder carcinoma, MNU-induced carcinoma | rat | D + Millipore support | Waymouth's MB 752/1 + S (10%) + AA + HC | 5 | 2–8 weeks | 272 |
| *Uterus* | | | | | | |
| Leiomyosarcoma | human | D + Millipore support | Eagle's + S (10%) | 10 | 11 days | 201 |
| Epithelioma (T8) | rat | B | BSS + S (20%) + CEE + mesonephros | NS | 30 days | 274 |

## References

1 Algard, F.T.: Hormone-induced tumours. I. Hamster flank-organ and kidney tumours in vitro. J. natn. Cancer Inst. *25:* 557–571 (1960).
2 Algard, F.T.: Hormone-induced tumours. II. Flank-organ epithelioma of the Syrian hamster in vitro. J. natn. Cancer Inst. *27:* 1493–1502 (1961).
3 Ambrose, E.J.; Easty, D.M.: Time-lapse filming of cellular interactions in organ culture. II. Behaviour of malignant cells. Differentiation *1:* 277–284 (1973).
4 Ambrose, E.J.; Easty, D.M.: Cellular dynamics of human breast carcinoma; in Dendy, Human tumors in short term culture, pp. 45-54 (Academic Press, London 1976).
5 Archer, F.L.: Normal and neoplastic human tissue in organ culture. Archs Path. *85:* 62–71 (1968).
6 Armstrong, R.C.; Rosenau, W.: Cocultivation of human primary breast carcinomas and embryonic mesenchyme resulting in growth and maintenance of tumour cells. Cancer Res. *38:* 894–900 (1978).
7 Aspegren, K.: 7,12-DMBA induced rat mammary tumour studied for hormonal responsiveness in vitro. 2. Organ cultures. Acta pathol. microbiol. scand., A, Pathol. *83:* 37–50 (1975).
8 Aspegren, K.: Hormone effects on human mammary cancer in organ cultures. Am. J. Surg. *131:* 575–580 (1976).
9 Aspegren, K.; Danielsson, H.: Growth quantitation of human mammary carcinoma in organ tissue culture. Am. J. Surg. *128:* 42–48 (1974).
10 Astedt, B.; Lundgren, E.; Roos, G.; Abu Sinna, G.: Release of various molecular forms of plasminogen activators during culture of human ovarian tumours. Thromb. Res. *13:* 1031–1037 (1978).
11 Auersperg, N.; Finnegan, C.V.: The differentiation and organization of tumours in vitro; in Sherbet, Neoplasia and cell differentiation, pp. 279–318 (Karger, Basel 1974).
12 Balls, M.; Monnickendam, M.: Organ culture in biomedical research (Cambridge University Press, Cambridge 1976).
13 Bang, F.B.; Niven, J.S.F.: A study of infection in organized tissue culture. Br. J. exp. Path. *39:* 317–322 (1958).
14 Beaupain, R.: Survie d'une tumeur maintenue en culture organotypique après irradiation aux rayons X. C.r. hebd. Séanc. Acad. Sci., Paris *271:* 2399–2402 (1970).
15 Beaupain, R.: Effets des rayons X sur l'incorporation de la thymidine tritiée dans les nodules cancéreux humain maintenus en culture organtypique de longue durée. C.r. hebd. Séanc. Acad. Sci., Paris *274:* 588–591 (1972).
16 Beaupain, R.: Action de doses répétées de rayons X sur la survie et la synthèse de l'ADN des explants cancéreux humains maintenus en culture organtypique de longue durée. C.r. hebd. Séanc. Acad. Sci., Paris *276:* 1925–1938 (1973).
17 Beaupain, R.; Wolff, E. Etude histochimique du mucus sécrété par une tumeur maligne du côlon cultivée in vitro. C. r. hebd. Séanc. Acad. Sci., Paris *264:* 1953–1956 (1967).
18 Beebe, S.P.; Ewing, J.: A study of the biology of tumour cells. Br. med. J. *ii:* 1559 (1906).
19 Beeby, D.I.; Easty, G.C.; Gazet, J.C.; Grigor, K.; Neville, A.M.: An assessment of the effects of hormones on short term organ cultures of human breast carcinomata. Br. J. Cancer *31:* 317–328 (1975).
20 Beeby, D.I.; Easty, G.C.; Gazet, J.C.; Neville, A.M.: Selection of patients with carci-

noma of the breast for endocrine therapy; in Dendy, Human tumours in short term culture, pp. 239–244 (Academic Press, London 1976).
21 Bissell, M.G.; Rubinstein, L.J.; Bignami, A.; Herman, M.M.: Characteristics of the rat C-6 glioma maintained in organ culture systems. Production of glial fibrillary acidic protein in the absence of gliofibrillogenesis. Brain Res. *82:* 77–89 (1974).
22 Braun, A.C.: Differentiation and dedifferentiation; in Becker, Cancer: a comprehensive treatise, vol. 3, pp. 3–20. (Plenum Press, New York 1975)
23 Breborowicz, J.; Easty, G.C.; Birbeck, M., Robertson, D.; Nery, R.; Neville, A.M.: The monolayer and organ culture of human colorectal carcinomata and the associated 'normal' colonic mucosa and their production of carcinoembryonic antigens. Br. J. Cancer *31:* 559–569 (1975).
24 Brinster, R.L.: Cultivation of the mammalian embryo; in Rothblat, Cristofalo, Growth, nutrition and metabolism of cells in culture, vol. 2, pp. 251–286 (Academic Press, New York 1972).
25 Brockas, A.J.; Wiernik, G.: Isolation of labelled glycoproteins from organ cultured carcinoma of the cervix; in Dendy, Human tumours in short term culture, pp. 270–276 (Academic Press, New York 1976).
26 Burtin, P.; Buffe, D.; Von Kleist, S.; Wolff, E.; Wolff, E.T.: Mis en évidence de l'antigène carcinoembryonnaire spécifique des cancers digestifs dans des tumeurs humaines entretenues en culture organotypique. Int. J. Cancer *5:* 88–95 (1970).
27 Cairns, J.: Mutation selection and the natural history of cancer. Nature, Lond. *255:* 197–200 (1975).
28 Carrel, A.; Burrows, M.T. Cultivation in vitro of malignant tumours. J. exp. Med. *13:* 571 (1911).
29 Chen, J.M.: The cultivation in fluid medium of organized liver, pancreas and other tissues of foetal rat. Expl. Cell Res. *7:* 518–529 (1954).
30 Cline, M.J.; Haskell, C.M.: Cancer chemotherapy; 3rd ed. (Saunders, Philadelphia 1980).
31 Cooper, P.; Golding, I.P.: Organ culture studies of human normal tissues and tumours. Acta Un. int. Cancr. *20:* 1288–1291 (1964).
32 Crocker, J.F.S.; Vernier, R.L.: Congenital nephroma of infancy: induction of renal structures by organ culture. J. Pediat. *80:* 69–73 (1972).
33 Cudennec, C.A.; Nicolas, J.-F.: Blood formation in a clonal cell line of mouse teratocarcinoma. J. Embryol exp. Morph. *38:* 203–210 (1977).
34 Dawe, C.J.: Cell sensitivity and specificity of response to polyoma virus. Natn. Cancer Inst. Monogr. *4:* 67–128 (1960).
35 Dawe, C.J.: Changes in cell interrelationships during epithelial carcinogenesis; in Nakahara, Takayama, Sugimura, Odashira, Topics in chemical carcinogenesis, pp. 401–427 (University Park Press, Tokyo 1972).
36 Dawson, M.: In vitro systems in basic biomedical research; in The future of animals, cells, models and systems in research, development, education and testing, pp. 185–215 (National Academy Science, Washington 1977).
37 DeCosse, J.J.; Gossens, C.L.; Kuzma, J.F.; Unsworth, B.R.: Breast cancer: induction of differentiation by embryonic tissue. Science, N.Y. *181:* 1057–1058 (1973).
38 DeCosse, J.J.; Gossens, C.L.; Kuzma, J.F.; Unsworth, B.R.: Embryonic inductive tissues that cause histologic differentiation of murine mammary carcinoma in vitro. J. natn. Cancer. Inst. *54:* 913–922 (1975).

39  De Grouchy, J.; Wolff, E.: Analyse chromosomique d'une tumeur cancereuse humaine en culture organotypique. Eur. J. Cancer *5:* 159–163 (1969).
40  Dendy, P.P.: The effects of X-rays on a human tumour growing in organotypic culture. Eur. J. Cancer *4:* 163–172 (1968).
41  Dendy, P.P.: Human tumours in short term culture (Academic Press, London 1976).
42  DeRidder, L.; Mareel, M.; Vakaet, L.: Adhesion of malignant and nonmalignant cells to cultured embryonic substrates. Cancer Res. *35:* 3164–3171 (1975).
43  DeRidder, L.; Mareel, M.; Vakaet, L.: Invasion of malignant cells into cultured embryonic substrates. Arch. Geschwulstforsch. *41:* 7–27 (1977).
44  Dermer, G.B.; Sherwin, R.P.: Autoradiographic localization of glycoprotein in human breast cancer cells maintained in organ culture after incubation with [$^3$H] fucose or [$^3$H] glucosamine. Cancer Res. *35:* 63–67 1975).
45  Dexter, D.L.; Kowalski, H.M.; Blazar, B.A.; Fligiel, Z.; Vogel, R.; Heppner, G.H.: Heterogeneity of tumour cells from a single mouse mammary tumour. Cancer Res. *38:* 3174–3181 (1978).
46  Dickson, J.A.; Suzangar, M.: In vitro sensitivity testing of human tumour slices to chemotherapeutic agents – its place in cancer therapy; in Dendy, Human tumours in short term culture, pp. 108–138 (Academic Press, London 1976).
47  Dickson, J.A.; Suzangar, M.: The in vitro response of human tumours to cytotoxic drugs and hyperthermia (42 °C) and its relevance to clinical oncology; in Balls, Monnickendam, Organ culture in biomedical research, pp. 417–461 (Academic Press, London 1976).
48  Dresden, M.H.; Heilman, S.A.; Schmidt, J.D.: Collagenolytic enzymes in human neoplasms. Cancer Res. *32:* 993–996 (1972).
49  Easty, D.M.; Easty, G.C.: An in vitro model for studying cell invasiveness; in Balls, Monnickendam, Organ culture in biomedical research, pp. 379–392 (Academic Press, London 1976).
50  Easty, D.M.; Wylie, J.A.H.: Screening of 12 gliomata against chemotherapeutic agents in vitro. Br. med. J. *i:* 1589–1592 (1963).
51  Easty, G.C.: Organ culture methods. Meth. Cancer Res. *5:* 1–43 (1970).
52  Easty, G.C.: Invasion by cancer cells; in Ambrose, Roe, Biology of cancer (Hooward, Chichester 1975).
53  Easty, G.C.; Easty, D.M.: An organ culture system for the examination of tumour invasion. Nature, Lond. *199:* 1104–1105 (1963).
54  Easty, G.C.; Easty, D.M.: Measurement of the ability of cells to infiltrate normal tissues in vitro. Br. J. Cancer *29:* 36–49 (1974).
55  Easty, G.C.; Powles, T.; Easty, D.M.; Dowsett, M.; Neville, A.: The detection of osteolytic substances produced by human breast tumour; in Balls, Monnickendam, Organ culture in biomedical research, pp. 367–377 (Cambridge University Press, Cambridge 1976).
56  Elias, J.J.; Armstrong, R.C.: Hyperplastic and metaplastic responses of human mammary fibroadenomas and dysplasias in organ culture. J. natn. Cancer Inst. *51:* 1341–1343 (1973).
57  Elias, J.J.; Rivera, E.: Comparison of the responses of normal, precancerous and neoplastic mouse mammary tissues to hormones in vitro. Cancer Res. *19:* 505–511 (1959).

58  Ellison, M.I.; Ambrose, E.J.; Easty, G.C.: Differentiation in a transplantable rat tumour maintained in organ culture. Expl. Cell Res. *55:* 198–204 (1969).
59  Evans, M.J.: The isolation and properties of a clonal tissue culture strain of pluripotent mouse teratoma cells. J. Embryol. exp. Morph. *28:* 163–176 (1972).
60  Feind, C.R.; Weber, C.J.; Derenoncourt, F.; Williams, G.A.; Hardy, M.A.; Reemtsma, K.: Survival and allotransplantation of cultured human parathyroids. Transplant. Proc. *11:* 1011–1016 (1979).
61  Fell, H.B.: The development of organ culture; in Balls, Monnickendam, Organ culture in biomedical research, pp. 1–13 (Cambridge University Press, Cambridge 1976).
62  Fell, H.B.; Robison, R.: The growth, development and phosphatase activity of embryonic avian femora and limb buds cultivated in vitro. Biol. J. *23:* 767–785 (1929).
63  Feng, J.; Melcher, A.H.; Brunette, D.M.; Moe, H.K.: Determination of *L*-ascorbic acid levels in culture medium: concentrations in commercial media and maintenance of levels under conditions of organ culture. In vitro *13:* 91–99 (1977).
64  Fidler, I.J.: Mechanisms of cancer invasion and metastasis; in Becker, Cancer: a comprehensive treatise, vol. 4, pp. 101–131 (Plenum Press, New York 1975).
65  Fidler, I.J.: Tumor heterogeneity and the biology of cancer invasion and metastasis. Cancer Res. *38:* 2651–2660 (1978).
66  Fidler, I.J.: General considerations for studies of experimental cancer metastasis. Meth. Cancer Res. *15:* 399–439 (1978).
67  Finkelstein, M.; Geier, A.; Horn, H.; Levij, I.S.; Ever-Hadni, P.: Effect of testosterone and estradiol-17β on synthesis of DNA, RNA and protein in human breast in organ culture. Int. J. Cancer *15:* 78–90 (1975).
68  Flaxman, B.A.; Chopra, D.P.; Newman, D.: Growth of mouse vaginal epithelial cells in vitro. In vitro *9:* 194–201 (1973).
69  Fogel, M.; Gorelik, E.; Segal, S.; Feldman, M.: Differences in cell surface antigens of tumour metastases and those of the local tumour. J. natn. Cancer Inst. *62:* 585–588 (1979).
70  Folkman, J.: The intestine as an organ culture; in Burdette, Carcinoma of the colon and antecedent epithelium, pp. 113–217 (Thomas, Springfield 1970).
71  Folkman, J.: Tumour angiogenesis; in Becker, Cancer: a comprehensive treatise, vol. 3, pp. 355–388 (Plenum Press, New York 1975).
72  Folkman, J.; Cotran, R.: Relation of vascular proliferation to tumour growth. Int. Rev. exp. Path. *16:* 207–248 (1976).
73  Folkman, J.; Greenspan, H.P.: Influence of geometry on control of cell growth. Biochim. biophys. Acta *417:* 211–236 (1975).
74  Folkman, J.; Long, D.M.; Becker, F.F.: Growth and metastasis of tumours in organ culture. Cancer *16:* 453–467 (1963).
75  Folkman, J.; Moscona, A.: Role of cell shape in growth control. Nature, Lond. *273:* 345–349 (1978).
76  Foulds, L.: Neoplastic development, vol. I (Academic Press, London 1969).
77  Franks, L.M.: Summary and future developments; in Balls, Monnickendam, Organ culture in biomedical research, pp. 549–556 (Cambridge University Press, Cambridge 1976).
78  Franks, L.M.: Primary cultures of human prostate. Meth. Cell Biol. *21B:* 153–169 (1979).

79  Franks, L.M.; Barton, A.A.: The effects of testosterone on the ultrastructure of the mouse prostate in vivo and in organ cultures. Expl Cell Res. *19:* 35–50 (1960).
80  Franks, L.M.; Riddle, P.N.; Carbonell, A.W.; Gey, G.O.: A comparative study of the ultrastructure and lack of growth capacity of adult human prostate epithelium. J. Path. *100:* 113–119 (1970).
81  Freshney, R.I.; Paul, J.: Culture of human tumour biopsies for the assessment of drug sensitivity; in Jacoby, Rajan, Tissue culture in medical research, pp. 179–188 (1974).
82  Geier, A.; Horn, H.; Levij, I.S.; Lichtshtein, E.; Finkelstein, M.: The metabolism of $^3$H-estradiol-17β in human breast cancer in organ culture. Eur. J. Cancer *11:* 127–130 (1975).
83  Geier, A.; Horn, H.; Lichtshtein, E.; Levij, I.S.; Finkelstein, M: The uptake and metabolism of [4-$^{14}$C] testosterone in human breast cancer grown in organ culture. Acta Endocr. *75:* 195–204 (1974).
84  Gewant, W.C.; Chasin, L.; Tilson, T.D.; Rutledge, C.; Goldenberb, I.S.: Lymph-node breast carcinoma interrelations in tissue culture. Surgery Gynec. Obstet. *133:* 959–962 (1971).
85  Grobstein, C.: Mechanisms of organogenetic tissue interaction. Natn. Cancer Inst. Monogr. *26:* 279–299 (1967).
86  Guillino, P.M.: Extracellular compartments of solid tumours; in Becker, Cancer: a comprehensive treatise, vol. 3, pp. 327–354 (Plenum Press, New York, 1975).
87  Harbitz, T.B.: Organ culture of benign nodular hyperplasia of human prostate in chemically-defined medium. Scand. J. Urol. *7:* 6–13 (1973).
88  Harbitz, T.B.; Falkanger, B.; Sander, S.: Benign prostatic hyperplasia of the human prostate exposed to steroid hormones in organ culture. Acta pathol. microbiol. scand., A., Pathol. *248:* suppl., pp. 89–93 (1974).
89  Hart, I.R.; Fidler, I.J.: An in vitro quantitative assay for tumour cell invasion. Cancer Res. *38:* 3218–3224 (1978).
90  Hecker, D. von; Deutschmann, A.: Histochemical and ultrastructural investigations on organ culture of malignant tumours. Acta histochem. *55:* 8–13 (1976).
91  Heckman, C.A.; Marchok, A.C.; Nettesheim, P.: Respiratory tract epithelium in primary culture: concurrent growth and differentiation during establishment. J. Cell Sci. *32:* 269–291 (1978).
92  Hegre, O.D.; McEnvoy, R.C.; Bachelder, V.; Lazarow, A.: Organ culture of fetal rat pancreas: quantitative analysis by linear scanning of islet and other tissue components. in vitro *7:* 366–376 (1972).
93  Herrmann, H.: Direct metabolic interaction between animal cells. Science, N.Y. *132:* 529–532 (1960).
94  Herwig, K.R.; Fischer, T.V.; Burkel, W.E.; Kahn, R.H.: Organ culture of canine prostate. Investve Urol. *15:* 291–294 (1978).
95  Heuson, J.C.; Coune, A.; Heimann, R.: Cell proliferation induced by insulin in organ culture of rat mammary carcinoma. Expl Cell Res. *45:* 351–360 (1967).
96  Heuson, J.C.; Legros, N.: Study of the growth-promoting effect of insulin in relation to carbohydrate metabolism in organ culture of rat mammary carcinoma. Eur. J. Cancer *4:* 1–7 (1968).
97  Heuson, J.C.; Legros, N.: Effect of insulin on DNA synthesis and DNA polymerase activity in organ culture of rat mammary carcinoma, and the influence of insulin pretreatment and of alloxan diabetes. Cancer Res. *31:* 59–65 (1971).

98   Heuson, J.C.; Legros, N.; Heuson-Stiennon, J.; Leclercq, G.; Pasteels, J.L.: Hormone dependency of rat mammary tumours; in Heuson, Maltheiem, Rozencweig, Breast cancer: trends in research and treatment, pp. 81–93 (Raven Press, New York 1976).
99   Heuson, J.C.; Pasteels, J.L.; Legros, N.; Heuson-Stiennon, J.; Leclercq, G.: Estradiol-dependent collagenolytic enzyme activity in long-term organ culture of human breast cancer. Cancer Res. *35:* 2039–2048 (1975).
100  Hill, B.T.; Baserga, R.: The cell cycle and its significance for cancer treatment. Cancer Treat. Rev. *2:* 159–175 (1975).
101  Hodges, G.M.: 'Adaptation' of a strain of neurovaccinia and of dermovaccinia to organ cultures of chick embryo skin. Acta virol. *10:* 450–454 (1966).
102  Hodges, G.M.: Influence of tissue differentiation on susceptibility to virus infection. J. gen. Virol. *1:* 487–494 (1967).
103  Hodges, G.M.: A review of methodology in organ culture; in Balls, Monnickendam, Organ culture in biomedical research, pp. 15–59 (Cambridge University Press, Cambridge 1976).
104  Hodges, G.M.: An overview of tissue culture procedures in tumour biopsy studies; in Dendy, Human tumours in short term culture, pp. 4–15 (Academic Press, London 1976).
105  Hodges, G.M.: Normal and neoplastic urothelium of human bladder in vivo and in vitro – an assessment of SEM studies. SEM Symp. *11:* 983–990 (1978).
106  Hodges, G.M.; Hallowes, R.C.: Biomedical research applications of scanning electron microscopy, vol. 1 (Academic Press, London 1979).
107  Hodges, G.M.; Hallowes, R.C.: Biomedical research applications of scanning electron microscopy, vol. 2 (Academic Press, London 1980).
108  Hodges, G.M.; Hicks, R.M.; Spacey, G.D.: Epithelial-stromal interactions in normal and chemical carcinogen-treated adult bladder. Cancer Res. *37:* 3720–3730 (1977).
109  Hoorn, B.; Tyrrell, D.A.J.: On the growth of certain 'newer' respiratory viruses in organ culture. Br. J. exp. Path. *46:* 109–118 (1965).
110  Hoorn, B.; Tyrrell, D.A.J.: Organ cultures in virology. Prog. medical Virol. vol. 11, pp. 408–450 (Karger, Basel 1969).
111  Hustin, J.: Effect of protein hormones and steroids on tissue cultures of endometrial carcinoma. Br. J. Obstet. Gynaec. *82:* 493–500 (1975).
112  Indo, K.; Wilson, B.R.: Foetal rat keratimizing epidermal cells in culture. Effects of long-term treatment by benzo(a)pyrene on their growth characteristics. J. natn. Cancer Inst. *59:* 867–874 (1977).
113  Iogannsen, M.G.; Pliss, G.B.: The organic culture of the urinary bladder with a flowing system of nutrition. Arkh. Anat. Gistol. Embriol. *70:* 109–114 (1976).
114  Jacobelli, S.; Sica, G.; Ranelletti, F.; Barile, G.: An assessment of the effects of steroid hormones and antiestrogens on short-term organ culture of human endometrial carcinoma. Eur. J. Cancer *14:* 931–938 (1978).
115  Jepson, A.: An in vitro model of a keratinizing squamous epithelium. Scand. J. dent. Res. *82:* 144–146 (1974).
116  Kadohama, N.; Kirdani, R.Y; Murphy, G.P.; Sandberg, A.A.: 5α-Reductase as a target enzyme for anti-prostatic drugs in organ culture. Oncology *34:* 123–128 (1977).
117  Kahn, R.H.; Burkel, W.E.; Perry, V.P.: Homeostatic and mass culture technology. J. natn. Cancer Inst. *53:* 1471–1477 (1974).

118 Kalus, M.: Carcinoma and adenomatous polyps of the colon and rectum in biopsy and organ tissue culture. Cancer 30: 972–982 (1972).
119 Kalus, M.; Ghidoni, J.J.; O'Neal, R.M.: The growth of tumours in matrix cultures. Cancer 22: 507–516 (1968).
120 Kalus, M.; O'Neal, R.M.: Organ tissue culture on a three-dimensional matrix of human fibrinfoam. Archs Path. 86: 52–59 (1968).
121 Karasek, M.A.: In vitro growth and maturation of epithelial cells from postembryonic skin. J. invest. Derm. 65: 60–66 (1975).
122 Keay, L.: The cultivation of animal cells and production of viruses in serum-free systems. Meth. Cell Biol. 20: 169–209 (1978).
123 Kiricuta, I.; Todorutiu, C.; Mulea, R.; Risca, R.: Axillary lymph-node and breast carcinoma interrelations in organ culture. Cancer 42: 2710–2715 (1978).
124 Kleinberg, D.L.: Human α-lactalbumin: measurement in serum and in breast cancer organ culture by radioimmunoassay. Science, N.Y. 190: 277–278 (1975).
125 Knazek, R.A.; Guillino, P.M.; Kohler, P.O.: Cell culture on artificial capillaries: an approach to tissue growth in vitro. Science, N.Y. 178: 65–66 (1972).
126 Kohorn, E.I.: The limitations of progesterone sensitivity testing of endometrial carcinoma using organ culture; in Dendy, Human tumours in short term culture, pp. 245–253 (Academic Press, New York 1976).
127 Kohorn, E.I.; Tchao, R.: The effect of hormones on endometrial carcinoma in organ culture. J. Obstet. Gynaec. Br. Commonw. 75: 1262–1267 (1968).
128 Kreisberg, J.I.; Brattain, M.G.; Pretlow, T.G.: Studies on human hyperplastic prostates maintained in organ culture. Investve. Urol. 15: 252–255 (1977).
129 Kruse, P.F., Jr.: Growth, nutrition and metabolism of cells in culture, in Rothblat, Cristofalo, Use of perfusion systems for growth of cell and tissue cultures, vol. 2, pp. 11–66 (Academic Press, New York 1972).
130 Lagios, M.D.: Hormonally enhanced proliferation of human breast cancer in organ culture. Oncology 29: 22–33 (1974).
131 Lakshmi, M.S.; Sherbet, G.V.: Embryonic and tumour cell interaction; in Sherbet, Neoplasia and cell differentiation, pp. 380–399 (Karger, Basel 1974).
132 Lambert, R.A.; Hanes, F.M.: Characteristics of growth of sarcoma and carcinoma cultivated in vitro. J. exp. Med. 13: 495 (1911).
133 Lasfargues, E.Y.: Cultivation and behaviour in vitro of the normal mammary epithelium of the adult mouse. Expl Cell Res. 13: 553–562 (1957).
134 Lasnitzki, I.: NIH Workshop on Kidney and Urology Program, in Grayhack, Wilson, Scherbenska Human benign prostatic hyperplasia in cell and organ culture, pp. 235–248 (1975).
135 Lasnitzki, I.: Growth and hormonal response of prostatic tumours; in Tannenbaum, Urologic pathology: the prostate, pp. 215–222 (Lea & Febiger, Philadelphia 1977).
136 Lasnitzki, I.; Whittaker, R.H. Withycombe, J.F.R.: The effect of steroid hormones on the growth pattern and RNA synthesis in human benign prostatic hyperplasia in organ culture. Br. J. Cancer 32: 168–178 (1975).
137 Leighton, J.: A sponge matrix method for tissue culture. Formation of organized aggregates of cells in vitro. J. natn. Cancer Inst. 12: 545–562 (1951).
138 Leighton, J.: Bioassay of cancer in matrix tissue culture systems; in Frei, The proliferation and spread of neoplastic cells. 21st Annu. Symp. on fundamental Cancer Research, Houston 1967, pp. 533–553.

139 Leighton, J.: Invasive growth and metastatis in tissue culture systems. Meth. Cancer Res. 4: 85–124 (1968).
140 Leighton, J.: Propagation of Cancer : targets for future chemotherapy. Cancer Res. 29: 2457–2465 (1969).
141 Leighton, J.: Cell propagation on miscellaneous culture supports: collagen-coated cellulose sponge; in Kruse, Patterson, Tissue culture methods and applications (Academic Press, New York 1973).
142 Leighton, J.; Abaza, N.; Tchao, R.; Geisinger, K.; Valentich, J.: Development of tissue culture procedures for predicting the individual risk of recurrence in bladder cancer. Cancer Res. 37: 2854–2859 (1977).
143 Leighton, J.; Justh, G.; Esper, M.: Collagen-coated cellulose sponge: three-dimensional matrix for tissue culture of Walker Tumour 256. Science N.Y. 155: 1259–1261 (1967).
144 Leighton, J.; Mark, R.; Justh, G.: Patterns of three-dimensional growth in vitro in collagen-coated cellulose sponge: carcinomas and embryonic tissues. Cancer Res. 28: 286–296 (1968).
145 Lustig, E.S.; De Mutrajt, H.A.: Acción del organizador primario del pono cultivado sobre el sarcoma. Revta Soc. argent. Biol. 37: 180–186 (1961).
146 Lustig, E.S.; Lustig, L.: Acción de inductors embrionarios sobre tejidos tumorales. Revta. Soc. argent. Biol. 40: 207–216 (1964).
147 Lustig, E.S.; Lustig, L.; Jauregui, H.: Action in vitro of the embryonic inducers on experimental and human tumours; in Katsuta, Cancer cells in culture, pp. 135–142 (University Park Press, Baltimore 1968).
148 Lyser, K.: Low- and high-voltage electron microscopy of a human neuroblastoma in long-term organ culture. Cancer Res. 34: 594–602 (1974).
149 Lyser, K.: The differentiation of glial cells and glial limitans in organ cultures of chick spinal cord. In Vitro 8: 77–84 (1972).
150 Lyser, K.: Organ culture of human nervous system tumours. In Vitro 12: 48–56 (1976).
151 Mandeville, R.; Holloway, A.; Lauchlan, S.C.; Simard, R.: Replication of herpes simplex virus type 2 in normal dysplastic and neoplastic human cervical epithelia. Eur. J. Cancer 15: 351–361 (1979).
152 Mareel, M.; De Brabander, M.: Effect of microtubule inhibitors on malignant invasion in vitro. J. natn. Cancer Inst. 61: 787–792 (1978).
153 Mareel, M.; De Bruyne, G.; DeRidder, L.: Invasion of malignant cells into $^{51}$Cr-labelled host tissues in organotypical culture. Oncology 34: 6–9 (1977).
154 Mareel, M.; Vakaet, L.; DeRidder, L.: Behaviour of Hela cells grown on young chick blastoderma in vitro. Virchows Arch. Abt. B Zellpath. 5: 277–278 (1970).
155 Mareel, M.; Vakaet, L.; DeRidder, L.: Possibility of distinction between malignant and nonmalignant cells by transplantation into chick blastoderma – further evidence from animal and human biopsy specimens. J. natn. Cancer Inst. 53: 1351–1358 (1974).
156 Mason, J.; Wolff, E.; Wolff, E.: Etude quantitative de la croissance organotypique in vitro d'une tumeur maligne humaine. C.r. hebd. Séanc. Acad. Sci., Paris 269: 1913–1917 (1969).
157 Masters, J.R.W.; Krishnaswamy, A.; Rigby, C.C.; O'Donoghue, E.P.N.: Quantitative organ culture: an approach to prediction of tumour response. Br. J. Cancer 41: suppl. IV, pp. 560–563 (1980).
158 Matoska, J.: Effect of progesterone on endometrial cancer in the organ culture. Neoplasma 17: 525–533 (1970).

159 Matoska, J.; Siracký, J.: Histology and ultrastructure of human breast cancer in organ culture. Neoplasma *21:* 685–696 (1974).
160 Matoska, J.; Stricker, F.: Following human tumours in primary organ culture. Neoplasma *14:* 507–519 (1967).
161 Matthias, M.: Methoden und Anwendungsmöglichkeiten der Organkultur in der klinischen und experimentellen Krebsforschung. Arch. Geschwulstforsch. *41:* 382–397 (1973).
162 McAteer, J.A.; Hegre, O.D.: Continuous flow method of organ culture. In Vitro *14:* 795–803 (1978).
163 McGowan, T.R.: Long-term support of intact skin in organ culture, with application to study of virus infection. Natn. Cancer Inst. Monogr. *4:* 95–115 (1963).
164 McLimans, W.F.: The gaseous environment of the mammalian cell in culture; in Rothblat, Cristofalo, Growth, nutrition, and metabolism of cells in culture, vol. I pp. 137–170 (Academic Press, New York 1972).
165 McMahon, M.J.; Butler, A.V.J.; Thomas, G.H.: Morphological responses of prostatic carcinoma to testosterone in organ culture. Br. J. Cancer *26:* 388–394 (1972).
166 McMahon, M.J.; Thomas, G.H.: The metabolism of testosterone by human prostate in organ culture. J. Endocr. *48:* XX–XXI (1970).
167 McMahon, M.J.; Thomas, G.H.: Morphological changes of benign prostatic hyperplasia in culture. Br. J. Cancer *27:* 323–335 (1973).
168 McRae, C.U.; Ghanadian, R.; Fotherby, K.; Chisholm, G.D.; The effect of testosterone on the human prostate in organ culture. Urology *45:* 156–162 (1973).
169 Medina, D. Tumour progression; in Becker, Cancer: a comprehensive treatise, vol. 3, pp. 99–119 (Plenum Press, New York 1975).
170 Monnickendam, M.A.; Balls, M.: Amphibian organ culture. Experientia *29:* 1–17 (1973).
171 Moscona, A.A.: Recombination of dissociated cells and the development of cell aggregates; in Willmer, Cells and tissues in culture, vol. I, pp. 489–529 (Academic Press, New York 1965).
172 Murrell, L.R.: Circumfusion organ culture of rat acinar pancreas with chemically defined medium; in Balls, Monnickendam, Organ culture in biomedical research, pp. 273–292 (Academic Press, London 1976).
173 New, D.A.T.: Whole-embryo culture and the study of mammalian embryos during organogenesis. Biol. Rev. *53:* 81–122 (1978).
174 Nicol, A.G.; Beck, J.S.: Persistence of an organ-specific antigen in organ and tissue cultures of hyperplastic human thyroid gland. Nature, Lond. *210:* 1227–1229 (1966).
175 Nicol, A.G.; Beck, J.S.: Organ culture of pathological human thyroid gland tissue. Br. J. exp. Path. *49:* 421–430 (1968).
176 Nissen, E.; Tanneberger, S.: Mixed organ culture as a tool for considering cellular tumour host relationships in tumour patients. Arch. Geschwulstforsch. *46:* 281–293 (1976).
177 Nissen, E.; Tanneberger, S.; Projan, A. Morack, G.; Peek, U.: Recent results of in vitro drug prediction in human tumour chemotherapy. Arch. Geschwulstforsch. *48:* 667–672 (1978).
178 Noguichi, P.D.; Johnson, J.B.; O'Donnell, R.; Pettricciani, J.C.: Chick embryonic skin as a rapid organ culture assay for cellular neoplasia. Science, N.Y. *199:* 980–983 (1978).

179 Nordqvist, R.S.: Hormone effects on carcinoma of the human uterine body studied in organ culture. Acta obstet. gynec. scand. *43:* 296–307 (1964).
180 Nordqvist, R.S.: Hormonal effects on endometrial carcinoma in vitro J. Endocr. *48:* 29–30 (1970).
181 Pasteels, J.L. Heuson, J.C.; Heuson-Stiennon, J.; Legros, N.: Effects of insulin, prolactin, progesterone, and estradiol on DNA synthesis in organ culture of 7,12-dimethylbenz(a)anthracene-induced rat mammary tumours. Cancer Res. *36:* 2162–2170 (1976).
182 Pasteels, J.L.; Heuson-Stiennon, J.A.; Legros, N.; Leclercq, G.; Heuson, J.C.: Organ culture of human breast cancer; in Heuson, Mattheiem, Rozencweig, Breast cancer: trends in research and treatment, pp. 141–150 Raven Press, New York (1976).
183 Paul, J.: Cell and tissue culture; 5th ed. (Livingstone, Edinburgh 1975).
184 Peillon, F.; Cesselin, F.; Garnier, P.E.; Brandi, A.M.; Donnadieu, M.; L'Hermite, M.; Dubois, M.P.: Prolactin secretion and synthesis in short- and long-term organ culture of pituitary tumours from acromegalic patients. Acta endocr. *87:* 701–715 (1978).
185 Peillon, F.; Gourmelen, M.; Donnadieu, M.; Brandi, A.; Sevaux, D.; Pham Hwu Trung, M.T.: Organ culture of human somatotrophic pituitary adenomas: ultrastructure and growth hormone production. Acta endocr. *79:* 217–229 (1975).
186 Picon, R.; Picon, L.; Chauffaux, S.; Ktorza, A.: Effects of canine fetal testes and testicular tumours on Mullerian ducts. Biol. Reprod. *18:* 459–467 (1978).
187 Pimm, M.V.; Baldwin, R.: Antigenic differences between methylcholanthrene-induced rat sarcomas and postsurgical recurrences. Int. J. Cancer *20:* 37–43 (1977).
188 Pontén, J.; MacIntyre, E.H.: Long-term culture of normal and neoplastic human glia. Acta path. microbiol. scand. *74:* 465–486 (1968).
189 Poole, A.R.: Tumour lysosomal enzymes and invasive growth; in Dingle, Lysosomes in biology and pathobiology, vol. 3, pp. 303–337 (North-Holland, Amsterdam 1973).
190 Poste, G.; Doll, J.; Hart, I.R.; Fidler, I.J.: In vitro selection of murine B16 melanoma variants with enhanced tissue-invasive properties. Cancer Res. *40:* 1636–1644 (1980).
191 Poste, G.; Fidler, I.J.: The pathogenesis of cancer metastasis. Nature, Lond. *283:* 139–146 (1980).
192 Pourreau-Schneider, N.: Action d'une tumeur de la glande interstitielle du testicule sur la prostate ventrale de rat en culture organotypique. C.r. Hebd. Séanc. Acad. Sci., Paris *270:* 242–244 (1970).
193 Pourreau-Schneider, N.; Felix, H.; Haemmerli, G.; Strauli, P.: The role of cellular locomotion in leukemic infiltration: an organ culture study on penetration of L 5222 rat leukemia cells into the chick embryo mesonephros. Virchows Arch. Abt. B. Zellpath. *23:* 257–264 (1977).
194 Prehn, R.T.: Neoplasia; in LaVia, Hill, Principles of pathobiology, pp. 191–241 (Oxford University Press, London 1972).
195 Rajewsky, M.F.: In vitro studies of cell proliferation in tumours. II. Characteristics of a standardized in vitro system for the measurement of $^3$H-thymidine incorporation into tissue explants. Eur. J. Cancer *1:* 281–287 (1965).
196 Reed, S.E.: Organ cultures for studies of viruses and mycoplasmas; In Balls, Monnickendam, Organ culture in biomedical research, pp. 515–532 (Cambridge University Press, London 1976).
197 Riches, A.C.; Littlewood, V.; Thomas, D.B.: The growth potential of tumour explants. J. Anat. *114:* 299–300 (1972).
198 Riches, A.C.; Shipman, P.A.; Littlewood, V.; Donaldson, I.; Thomas, G.H.: Organ

culture – a model for defining the hormone dependency of human benign prostatic hyperplasia; in Dendy, Human tumours in short term culture, pp. 227–238 (Academic Press, London 1976).
199   Rickenbacher, J.: Infektionsversuche mit Vakzinevirus an menschlicher embryonaler Haut in vitro. Pathol. Microbiol. 27: 624–637 (1964).
200   Rivera, E.M.; Elias, J.J.; Bern, H.A.; Napalkov, N.P.; Pitelka, D.R.: Toxic effects of steroid hormones on organ cultures of mouse mammary tumours, with a comment on the occurrence of viral inclusion bodies. J. natn. Cancer Inst. 31: 671–687 (1963).
201   Röller, M.R.; Owen, S.P.; Heidelberger, C.: Studies on the organ culture of human tumours. Cancer Res. 26: 626–637 (1966).
202   Rousseau, M.F.; Nabarra, B.; Nezelof, C.: Behaviour of Wilms tumour and normal metanephros in organ culture. Eur. J. Cancer 10: 461–466 (1974).
203   Rousseau-Merck, M.F.; Lombard, M.N.; Nezelof, C.; Mouly, H.: Limitation of the potentialities of nephroblastoma differentiation in vitro. Eur. J. Cancer 13: 163–170 (1977).
204   Rovin, S.: The influence of carbon dioxide on the cultivation of human neoplastic explants in vitro. Cancer Res. 22: 384–387 (1962).
205   Rubinstein, L.J.; Conley, F.K.; Herman, M.M.: Studies on experimental malignant nerve sheath tumours maintained in tissue and organ culture systems. I. Light microscopy observations. Acta neuropath. 34: 277–291 (1976).
206   Rubinstein, L.J.; Herman, M.M.: Studies on the differentiation of human and experimental gliomas in organ culture systems. Recent Results Cancer Res. 51: 35–51 (1975).
207   Rubinstein, L.J.; Herman, M.M.; Foley, V.L.: In vitro characteristics of human glioblastomas maintained in organ culture systems. Light microscope observations. Am. J. Path. 71: 61–80 (1973).
208   Rubinstein, L.J.; Herman, M.M.; Hanbery, J.W.; The relationship between differentiating medulloblastoma and dedifferentiating diffuse cerebellar astrocytoma: light, electron microscopic, tissue and organ culture observations. Cancer 33: 675–690 (1974).
209   Rumery, E.P.; Blandau, R.J.: Culture of mammalian embryonic ovaries and oviducts; in Daniel, Methods in mammalian embryology, pp. 472–495 (Freeman, San Francisco 1971).
210   Sakakura, T.; Sakagami, Y; Nishizuka, Y: Acceleration of mammary cancer development by grafting of fetal mammary mesenchymes in C3H mice. Gann 70: 450–466 (1979).
211   Saksela, E.; Holmstrom, T.; Grahne, B.: Growth pattern of meningiomas penetrating the skull base. Acta oto-lar. 74: 363–370 (1972).
212   Sasaki, K.: Measurement of tritiated thymidine labeling index by incubation in vitro of surgically removed cervical cancer. Gann 68: 307–313 (1977).
213   Saxen, L.; Vainio, T.; Toivonen, S.: Viral susceptibility and embryonic differentiation. 1. The histopathology of mouse kidney rudiments infected with polyoma and vesicular stomatitis virus in vitro. Acta path. nicrobiol. scand. 58: 191–211 (1963).
214   Schaeffer, W.I.: Proposed usage of animal tissue culture terms (revised 1978). Usage of vertebrate cell, tissue and organ culture terminology. In Vitro 15: 649–653 (1979).
215   Schleich, A.B.: Use of an organ culture system to demonstrate various types of malignant growth; in Dendy, Human tumours in short term culture, pp. 55–58 (Academic Press, New York 1976).
216   Schleich, A.B.; Frick, M.; Mayer, A.: The confrontation of normal tissue and malignant

cells in vitro – human decidua gravidatis and Hela cells. A model for studies on tumour invasion. Z. Krebsforsch. 82: 247–255 (1974).
217 Schleich, A.B.; Frick, M.; Mayer, A.: Patterns of invasive growth in vitro Human decidua graviditatis confronted with established human cells and primary human explants. J. natn. Cancer Inst. 56: 221–237 (1976).
218 Schrodt, G.R.; Foreman, C.D.: In vitro maintenance of human hyperplastic prostatic tissue. Investve Urol. 9: 85–94 (1971).
219 Schwartz, B.: Culture of the lens (33°); in Kruse, Patterson, Tissue culture methods and applications, pp. 305–306 (Academic Press, New York 1973).
220 Seilern-Aspang, F.; Honus, E.; Kratochwil, K.: Cartilage induction in a fowl sarcoma in vitro by the addition of chorda from chick embryos. Acta biol. med. germ. 10: 447–452 (1963).
221 Seilern-Aspang, F.; Honus, E.; Kratochwil, K.: Die Verknorpelung menschlicher Sarkome durch chondrogene Induktoren der Hühnerchorda. Acta biol. med. germ. 11: 281–285 (1963).
222 Shaffer, B.M. The culture of organs from the embryonic chick on cellulose-acetate fabric. Exptl. Cell Res. 11: 244–248 (1956).
223 Shipman, C.: Control of culture pH with synthetic buffers; in Kruse, Patterson, Tissue culture methods and applications, Academic Press, New York and London, pp. 709–713 (1973).
224 Sigot-Luizard, M-F.: The association of tumour and embryonic cells in vitro; in Sherbet, Neoplasia and cell differentiation, Karger, Basel. pp. 350–379 (1974).
225 Simnett, J.D.; Fisher, J.M.: An assay system for humoral growth factors. Natl. Cancer Inst. Monogr. 38: 19–28 (1973).
226 Sipe, J.C.; Herman, M.M.; Rubinstein, L.J.: Electron microscopic observations on human glioblastomas and astrocytomas maintained in organ culture systems. Am. J. Pathol. 73: 589–606 (1973).
227 Sipe, J.C., Rubinstein, L.J.; Herman, M.M.; Bignami, A.: Ethylnitrosurea-induced astrocytomas: Morphologic observations on rat tumours maintained in tissue and organ culture systems. Lab. Invest. 31: 571–579 (1975).
228 Siracký, J.: An approach to the problem of heterogeneity of human tumour-cell populations. Br. J. Cancer 39: 570–576 (1979).
229 Siracký, J.; Matoska, J.; Siracka, E.: Morphology and autoradiography studies of gynaecological tumours in organ culture. Neoplasma 21: 307–312 (1974).
230 Spence, A.M.; Rubinstein, L.J.: Cerebellar capillary hemangioblastoma. Its histogenesis studied by organ culture and electron microscopy. Cancer 35: 326–341 (1975).
231 Spratt, N.T., Jr.: A simple method for explanting and cultivating early embryos in vitro. Science, N.Y. 106: 452 (1947).
232 Steel, G.G.; Bensted, J.P.M.: In vitro studies of cell proliferation in tumours. 1. Critical appraisal of methods and theoretical considerations. Eur. J. Cancer 1: 275–279 (1965).
233 Stenbäck, F.: Characteristics of neoplastic progression and tumour behaviour in skin carcinogenesis by different methods. Oncology 37: 163–168 (1980).
234 Stinson, S.F.; Lilga, J.C.; Reese, D.H.; Friedman, R.D.; Sporn, M.B.: Quantitation with an automated image analyzer of nuclear-cytoplasmic changes induced by hydrocortisone in bladder epithelium. Cancer Res. 37: 1428–1431 (1977).
235 Stoll, B.A.: Investigation of organ culture as an aid to the hormonal management of breast cancer. Cancer 25: 1228–1233 (1970).

236 Strangeways, T.S.P.; Fell, H.P.: Experimental studies on the differentiation of embryonic tissues growing in vivo and in vitro. 1. The development of the undifferentiated limb bud. Proc. R. Soc. *99:* 340–364 (1926).
237 Takizawa, S.; Furth, J.J.; Furth, J.: DNA synthesis in autonomous and hormone-responsive mammary tumours. Cancer Res. *30:* 206–210 (1970).
238 Takizawa, S.; Furth, J.J.; Furth, J.: Biological and technical aspects of nucleic acid synthesis in cultures of mammary tumours. Cancer Res. *30:* 211–220 (1970).
239 Tanneberger, S.: Clinical cell biology and cancer. Arch. Geschwulstforsch. *39:* 44–47 (1972).
240 Tanneberger, S.; Mohr, A.: Biological characterization of human tumours by means of organ culture and individualized cytostatic cancer treatment. Arch. Geschwulstforsch. *42:* 307–315 (1973).
241 Tannock, I.F.: Cellular kinetics of experimental tumours in solid and ascitic form; in Burdette Carcinoma of the colon and antecedent epithelium, pp. 230–242 (Thomas, Springfield 1970).
242 Tarin, D.; Tissue interactions in carcinogenesis (Academic Press, New York 1972).
243 Tarin, D.: Fundamental aspects of metastasis, in Weiss, Cellular interactions in neoplasia, pp. 151–187 (North-Holland, Amsterdam 1976).
244 Taylor, W.G.: 'Feeding the baby' – serum and other supplements to chemically defined medium. J. natn. Cancer Inst. *53:* 1449–1457 (1974).
245 Tchao, R.; Easty, D.W.; Ambrose, E.J.: The use of $^{32}$P-labelled orthophosphate for the assay of chemotherapeutic agents on tumours maintained in organ culture. Br. J. Cancer *21:* 821–827 (1967).
246 Tchao, R.; Easty, G.C.; Ambrose, E.J.; Raven, R.W.; Bloom, H.J.G.: Effect of chemotherapeutic agents and hormones on organ cultures of human tumours. Eur. J. Cancer *4:* 39–44 (1968).
247 Thomas, J.A.: Organ culture (Academic Press, New York 1970).
248 Tobin, S.M.; Wilson, W.D.; Papsin, F.R.: Organ culture model for the study of HVH-II infections in carcinoma of the cervix. Obstet. Gynec. *53:* 559–564 (1979).
249 Trowell, D.A.: The culture of mature organs in a synthetic medium. Expl Cell Res. *16:* 118–147 (1959).
250 Tsypkin, L.B.; Voroshilova, M.K.; Goryunova, A.G.; Lavrova, I.K.: The results of organ cultivation of human gastric and rectal cancer tumours. Arckh. Patol. *35:* 25–31 (1973).
251 Tsypkin, L.B.; Voroshilova, M.K.; Goryunova, A.G.; Lavrova, I.K.; Koroleva, G.A.: The morphology of tumours of the human gastrointestinal tract in short-term organ culture and the reaction of these tumours to infection with poliovirus. Cancer *38:* 1796–1806 (1976).
252 VandenBerg, S.R.; Herman, M.M.; Ludwin, S.K.; Bignami, A.: An experimental mouse testicular teratoma as a model for neuroepithelial neoplasia and differentiation. Am. J. Path. *79:* 147–168 (1975).
253 Vera, J.; Licata, A.; Bartter, F.C.: A convenient, inexpensive organ culture device. Analyt. Biochem. *77:* 540–541 (1977).
254 Volpino, G.: Alcune esperienze sull cancro trapiantabule dei topi. Pathologice *2:* 495 (1910).
255 Voroshilova, M.K.; Goryunova, A.G.; Tsypkin, L.B.; Lavrova, I.K.; Koroleva, J.A.: Organ explant cultures of some human cancer tumours and the possibility of their use

for investigation of oncotropic properties of poliovirus. Acta. virol. *18:* 129–134 (1974).
256  Waymouth, C.: Osmolality of mammalian blood and of media for culture of mammalian cells. In Vitro *6:* 109–127 (1970).
257  Waymouth, C.: Growth, nutritition, and metabolism of cells in culture, in Rothblat, Cristofalo, Construction of tissue culture media, vol. I pp. 11–47 (Academic Press, New York 1972).
258  Waymouth, C: Tissue culture methods and applications, in Kruse, Patterson, Determination and survey of osmolality in culture media, pp. 703–709 (Academic Press, New York 1973).
259  Waymouth, C.: 'Feeding the baby' – designing the culture milieu to enhance cell stability. J. natn. Cancer Inst. *53:* 1443–1448 (1974).
260  Waymouth, C.: Nutritional requirements of cells in culture, with special reference to neural cells; in Fedoroff, Hertz, Cell, tissue and organ cultures in neurobiology, pp. 631–648 (Academic Press, New York 1977).
261  Wellings, S.R.; Jentoft, V.L.: Organ cultures of normal, dysplastic, hyperplastic and neoplastic human mammary tissues. J. natn. Cancer Inst. *49:* 329–338 (1972).
262  Welsch, C.W.: Prolactin and the development and progression of early neoplastic mammary gland lesions. Cancer Res. *38:* 4054–4058 (1978).
263  Welsch, C.W.; Calaf de Iturri; Brennan, M.J.: DNA synthesis of human, mouse and rat mammary carcinomas in vitro: influence of insulin and prolactin. Cancer *38:* 1272–1281 (1976).
264  Welsch, C.W.; Dombroske, S.E.; McManus, M.J.: Effects of insulin, human placental lactogen and human growth hormones on DNA synthesis in organ cultures of benign human breast tumours. Br. J. Cancer *38:* 258–261 (1978).
265  Welsch, C.W.; Dombroske, S.E.; McManus, M.J.; Calaf, G.: Effect of human, bovine and ovine prolactin on DNA synthesis by organ cultures of benign human breast tumours. Br. J. Cancer *40:* 866–871 (1979).
266  Welsch, C.W.; McManus, M.J.: Stimulation of DNA synthesis by human placental lactogen or insulin in organ cultures of benign human breast tumours. Cancer Res. *37:* 2257–2261 (1977).
267  Welsch, C.W.; Rivera, E.M.: Differential effects of estrogen and prolactin on DNA synthesis in organ cultures of DMBA-induced rat mammary carcinoma. Proc. Soc. exp. Biol. Med. *139:* 623–626 (1972).
268  Wessells, N.K.: Avian and mammalian organ culture; in Wilt, Wessells, Methods in developmental biology, pp. 445–456 (Crowell, New York 1967).
269  Wessells, N.K.: Tissue interactions and development (Benjamin, Menlo Park 1977).
270  Willcox, P.A.; Thomas, G.H.: Oestrogen metabolism in cultured human breast tumours. Br. J. Cancer *26:* 453–460 (1972).
271  Willmer, E.N.: Cells and tissue in culture (Academic Press, New York 1965).
272  Wilson, P.D.; Hodges, G.M.: Focal distribution of surface marker enzymes after long-term culture of adult rat bladder epithelium and methylnitrosourea (MNU)-induced bladder tumours. J. Histochem. Cytochem. *27:* 1236–1246 (1979).
273  Wilson, G.D.; Woods, K.L.; Walker, R.A.; Howell, A.: Effects of prolactin on lactalbumin production by normal and malignant human breast tissue in organ culture. Cancer Res. *40:* 486–489 (1980).
274  Wolff, E.: Organ chimeras and organ culture of malignant tumours; in Thomas, Organ culture, pp. 459–496 (Academic Press, New York 1970).

275 Wolff, E.: Techniques of organotypic culture; in Dendy, Human tumours in short term culture, pp. 40–44 (Academic Press, London 1976).
276 Wolff, E.; Goube, P.: L'application des cultures d'organes embryonnaires à la culture de virus pathogènes. C.r. hebd. Séanc. Acad. Sci., Paris 248: 490–492 (1959).
277 Wolff, E.; Goube, P.: Sur les lésions histopathologiques spécifiques provoquées par le virus de la variole aviare dans des explants de peau embryonnaire de poulet en culture d'organe in vitro. Annls Inst. Pasteur, Paris 98: 325–334 (1960).
278 Wolff, E.; Haffen, K.: Sur une methode de culture d'organes embryonnaires in vitro. Tex. Rep. Biol. Med. 10: 463–472 (1952).
279 Wolff, E.; Smith, J.; Wolff, E.: Organotypic cultures of human tumours; in Balls, Monnickendam, Organ culture in biomedical research, pp. 405–415 (Academic Press, London 1976).
280 Yarnell, M.; Ambrose, E.J.: Studies of tumour invasion in organ culture. I. Effects of basic polymers and dyes on invasion and dissemination. Eur. J. Cancer 5: 255–263 (1969).
281 Yarnell, M.; Ambrose, E.J.: Studies of tumour invasion in organ culture. II. Effects of enzyme treatment. Eur. J. Cancer 5: 265–269 (1969).
282 Yarnell, M.; Ambrose, E.J.; Shepley, K.; Tchao, R.: Drug assays on organ cultures of biopsies from human tumours. Br. med. J. ii: 490–491 (1964).
283 Yen, E.H.K.; Melcher, A.H.: Continuous-flow culture system for organ culture of large explants of adult tissue: effect of oxygen tension on mouse molar periodontium. In Vitro 14: 811–818 (1978).

G.M. Hodges, Dr. Sc., Imperial Cancer Research Fund, Lincoln's Inn Fields, London WC2A 3PX (England)

# Control of Proliferation of Normal and Neoplastic Cells in Culture

*P. A. Riley*

Department of Biochemical Pathology, University College School of Medicine, London, England

## I. Introduction

An understanding of the mechanisms that control the proliferation of cells in metaozoans is central to the systems of biological thought concerned with those major unsolved questions of biology – the nature of development and differentiation, ageing and cancer. A notable contribution to the study of this subject, particularly in recent years, has been made by investigations using cells in culture as experimental models of the processes of growth control in vivo. The extent to which the conditions existing in vitro resemble the circumstances experienced by cell populations in body tissues is debatable and, as will emerge from the subsequent discussion, there are many factors which exert a significant influence on cultured cells that do not readily find counterparts in the homeostatic environment of the body.

In particular the restrictions on the growth of cells imposed by the depletion of nutrients or growth factors or the accumulation of metabolic products in the pericellular fluid are likely to be of less significance in the body than under culture conditions. Conversely, the contribution of autoinhibitory controls to the regulation of population size is probably of greater importance to tissue homeostasis in vivo than in cultured cells.

In a topic beset, as this is, by caveats it must also be made clear that the behaviour of cultivated cells may not necessarily correspond to patterns observed in the body because the cultured cells become selected by successful proliferation under the artificial conditions into which they are introduced. The criteria of selection are, therefore, likely to differ significantly from those existing in vivo. However, despite this drawback, it is often possible to learn something about the general aspects of cellular behaviour from artificially selected cell lines, although it is not always possible to draw a correct analogy of in vivo behaviour from observations on in vitro characteristics since there may be no equivalent set of circumstances to those experienced by the cultured cells to be found in the body.

## II. Control of Proliferation

The concept of control implies some form of regulation involving a property of the controlled entity which results in an alteration of its behaviour. The generalised idea behind this is the feedback loop or closed loop control system. For control to be exercised the system must possess a mechanism for monitoring the action of a control element in such a way as to tend to reduce to zero the deviation of the signal from the desired level. The essential character of a controlling feedback loop is that it includes a sign reversal, i.e. it is a negative feedback loop.

Although negative feedback was an important feature of *James Watt's* steam 'governor', invented in 1788, the origins of negative feedback loops are much earlier in the Hellenistic period. *Ktesibios* (circa 250 BC) invented a water clock with a float regulator, *Philon* (circa 220 BC) devised a constant level device for oil lamps, and *Heron* (circa 50 AD) constructed several devices, such as float regulators and 'inexhaustible goblets' [157]. In physiology the regulatory concept was applied more than a century ago by *Bernard* [20] and subsequently by *Cannon* [39]. The general mathematical principles have been elaborated by *Wiener* [258]. The control of cell population growth in vivo has been ascribed in general terms to indirect feedback [33, 55] or a direct autoinhibitory influence [174, 192, 254].

In general terms the concept of regulation implies a deterministic approach to the control of cell growth. The main observational justification for a deterministic approach comes from the demonstration of clonal variations in growth behaviour of cells which are in identical environments. Analysis of the distribution of intermitotic intervals in cultures of guinea pig epithelial cells have shown that the proliferative behaviour of the cells is largely determined by inherited factors and independent of the microenvironmental conditions [199, 200]. Nevertheless, probabilistic models, such as that proposed by *Smith and Martin* [226] are useful in the analysis of the growth kinetics of polyclonal populations of cells.

## III. Hierarchical Controls

In relation to the control of proliferation in cells it is evident that many factors influence the rate of growth and that a hierarchical system of controls is involved.

The apex of this hierarchy is an intracellular regulatory mechanism which determines whether a paricular cell is going to prepare for mitosis or not. This mechanism I shall refer to as intrinsic regulation of proliferation (IRP). This intrinsic control is subject to modification by a number of extrinsic factors which operate in a permissive fashion. These extrinsic permissive factors may be divided into two classes. The first group includes microenvironmental conditions, such as availability of substratum and essential nutrients, or the concentration of metabolic products of cells. These factors act as controls of proliferation *within* a population or sub-population of cells and may be termed intercellular controls. The second category includes hormonal influences exerted either through hormones with a general action on cells, such as insulin, or growth factors or growth inhibitors of greater or lesser specificity. These are controlling agencies operating between different cell populations and might be termed interpopulation controls. It is probable that in vivo these controls are important in tissue homeostasis. This classification is summarised in table I. Superimposed on these physiological agencies are the effects of extraneous toxins and inhibitors. The following discussion will deal in turn with the intrinsic regulatory mechanism, the microenvironmental conditions which impose

*Table I.* Classification of proliferative control systems

1 Intrinsic regulation of cell proliferation
2 Extrinsic permissive factors
   a Intercellular controls (microenvironmental)
     i Space
     ii Nutrients
     iii Metabolic products

   b Interpopulation controls (hormones)
     i General hormones
     ii Growth factors
     iii Chalones

limitations on the capacity of the cell to proliferate, and hormones which are the means of communication between cell populations.

## IV. Intrinsic Regulation of Proliferation

### A. Cell Division

The evolutionary origin of cell division is obscure. It has been suggested that the process may have originated as a mechanism to reduce the size of unicellular organisms since with increasing volume the internal diffusion distances for metabolites and the comparative reduction of surface area are disadvantageous. In the antecambrian era, before the evolution of photosynthetic organisms, the anabolic activities of primitive cells were sustained by an external input of energy consisting essentially of the uptake of reducing substances. Any process leading to a diminution in the availability of electron donors from the environment would be detrimental to the metabolism of the cell. Elementary geometric considerations (taking a cell to be spherical and homogenous in composition) show that, as the radius increases, the steady-state concentration of a substance utilised at a rate proportional to the cell volume diffusing from the pericellular environment falls linearly with increasing cellular radii according to the expression:

$$[x] = K - k/3r, \tag{1}$$

where $[x]$ is the mean intracellular concentration of the substance; $K$ is the external concentration of the agent which is assumed to be constant; $k$ is a constant reflecting the ratio of the utilisation rate and the flux rate across the cell membrane; and $r$ is the radius of the cell (fig. 1). This reasoning indicates that there is an upper limit to the size of a cell and that continued efficiency of metabolism can be supported only by the intercession of a new strategy. Reduction of cytoplasmic mass by piecemeal losses is less efficient than cell division, of which the simplest mode is binary fission giving rise to two equivalent products. The importance of this proliferative stage in evolutionary history is manifest since it would enable selective forces to operate. Selection would tend to favour cells that divided rapidly, i.e. before the growth-limiting effects of size were a significant factor, and in which partition of informational components was equal.

It might, therefore, be anticipated that, in general, a balance would be

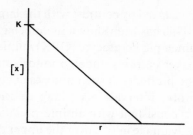

*Fig. 1.* Relationship between the cellular radius (r) and the mean intracellular concentration of a substance [x] diffusing into, and being utilised by, a cell.

struck between the size of the cellular genome and an optimum cell size. Certainly, there exists sufficient experimental evidence to indicate that eukaryotic cells can regulate the nuclear-cytoplasmic ratio so that, under balanced growth conditions, a constant average cell size is maintained over succeeding generations.

1. Constant Ratio of Genome to Cytoplasmic Mass

It has been proposed for some considerable time that cells maintain a constant relationship between the mass or volume of the cytoplasm and the mass of chromatin that they contain. *Yucas* et al. [265] showed that there is an approximately linear relationship between cell size and ploidy or nuclear size and a similar relationship between the number of nuclei per cell and cellular mass has been demonstrated for multinucleated cells [50, 124]. In prokaryotes a constant proportionality exists between cell mass and the number of initiation sites for DNA replication [67]. Evidence that this constant proportionality is subject to regulatory mechanisms comes from experiments in which the nuclear-cytoplasmic ratio is disturbed. If DNA synthesis is selectively inhibited without influencing the increase in cytoplasmic mass the nuclear-cytoplasmic ratio will be lowered and subsequent release from inhibition has been shown to result in compensatory increase in initiation of DNA replication in prokaryotes [186] and in shortened intermitotic periods in several eukaryote systems [28, 85, 165]. Conversely, if the cytoplasmic mass is reduced by amputation mitosis is delayed, as shown in amoeba by *Hartmann* [106] and *Prescott* [183, 184]. A natural perturbation to the nuclear-cytoplasmic ratio results if unequally sized cells are produced at cytokinesis. The smaller cell is equivalent to a cell in which a proportion of the cytoplasm has been amputated and the next division is correspondingly delayed by contrast with the larger cell to which the reverse would apply. This has been shown in amoeba by *Prescott* [183] and more recently in guinea pig keratocyte (GPK) cultures [201]. A proportion of cells in monolayer cultures undergo mitosis in a plane vertical to the substratum and the products of such mitoses exhibit transient variations in their intermitotic intervals according to their relative position. The observed effect is consistent with an interpretation based on the lowermost cell having an advantage in size over the upper cell after division in hypopyknic medium (see section V.A.1.a).

Reduction in the size of the genome by ultraviolet irradiation, as in experiments performed on syncytial slime molds [65, 216], results in shorter subsequent intermitotic intervals. Evidence that inhibition of protein synthesis, and hence inhibition of increase in cell mass, inhibits DNA synthesis has been reported by many workers [23, 128, 267]. *Terasima and Yasukawa* [237] showed that periods of inhibition of protein synthesis during G1 resulted in a proportional delay in the initiation of DNA synthesis in L cells. The possible significance of this relationship between inhibition of protein synthesis and replication of the genome has been discussed by *Mitchison* [164] particularly with regard to the dose responsiveness and lag period when RNA synthesis is inhibited [16, 41, 128]. Higher doses of actinomycin D than those necessary to inhibit progress to, and completion of, mitosis are necessary to inhibit initiation of DNA synthesis and a lag period of at least 2 h occurs [16, 236] and it seems probable that this is related to the synthesis of initiator proteins for DNA synthesis [19, 170].

## 2. Size Monitoring by Cells

The general inference from the evidence outlined above is that the cell possesses a mechanism for constantly monitoring the cytoplasmic mass or volume in relation to the size of the genome. Several possible mechanisms for such a process have been examined by *Fantes* et al. [73]. The most economical model, and the only system that is consistent with the compensation of perturbations in nuclear-cytoplasmic ratio without growth [80], involves a mechanism for sensing the concentration of a substance that is generated by the genome in a dose-dependent manner and is diluted in inverse proportion to the cytoplasmic volume. The rate of change in the amount of such a putative substance (y) may be described by:

$$dy/dt = aG - by, \qquad (2)$$

where $G$ represents the size of the genome and $a$ and $b$ are positive constants representing the rates of synthesis and degradation respectively. At the steady state ($dy/dt = 0$) the concentration of substance is given by:

$$[y] = \alpha\, G/V, \tag{3}$$

where $\alpha$ is the ratio of the rates of synthesis and degradation. Therefore, assuming that changes in the amount of the substance occur rapidly in comparison with alterations in the size of the cell or the genome, the intracellular concentration [y] will be proportional to the ratio of the size of the genome and the cellular volume (V). If there exists a mechanism that is responsive to the concentration of y then this could act as a monitoring system sensitive both to alterations in the size of the genome and in cytoplasmic volume. The substance proposed in this model has properties which would make a species of RNA that is synthesised in amounts proportional to the size of the genome and is taken up in proportion to the cytoplasmic mass, an attractive candidate for the role. Since there is evidence that the bulk of the ribosomal RNA is synthesised at differing rates according to the proliferative state of the cell [154] a possible candidate for the controlling substance might be a sub-species of ribosomal RNA such as 5S rRNA.

### B. Control Point

From the most general selective considerations it would be advantageous for cells to have developed a mechanism for ensuring that the genome is reduplicated prior to cellular fission and, although over a quarter of a century after the delineation of the phases of the eukaryotic cell cycle by *Howard and Pelc* [119] relatively little is known about the precise molecular events that regulate the progression of cells from one phase to the next, it is generally agreed that the control point in the proliferative process is at the boundary of the G1 and S phases. The importance of cell mass in determining this point in the cell cycle was inferred some years ago by *Killander and Zetterberg* [126, 127]. Cytochemical studies on mouse L cells in monolayer culture indicated that there was a significantly smaller variation in the mass of individual cells at the beginning of S phase than at the completion of mitosis. In addition, the variability of the time spent in G1 was much greater than the mass variation at the commencement of DNA synthesis and this was shown to be due to the compensatory lengthening of G1 in cells with low initial mass [127]. It is noteworthy that an inverse relationship was noted between the length of G1 and the S + G2 phases which has been confirmed in other studies [222] and suggests that general preparative syntheses take place during G1 that are not subject to control of the kind under discussion. Chromosome mapping data from the experiments of *Baserga* et al. [17] indicate that transcription of many parts of the genome occurs in a sequence commencing about 6 h before transition from G1 to S. The factors coded for by these sections of the genome probably include DNA polymerase [22] and strand-separating agents, as shown by the susceptibility of DNA to single-strand specific nuclease degradation for a considerable period preceding initiation of replication [53]. Other initiating factors are undoubtedly synthesised but at present it is not clear what these factors comprise. Their failure to be synthesised in non-proliferative cells, such as adult neurones, as shown by experiments on nuclear transplantation into *Xenopus* oocytes [96], attests to their fundamental importance in the proliferation of cells; but it would appear probable that many of the factors involved are prerequisites for DNA synthesis but have no direct controlling function as regulators of transition into S phase.

## C. Inducers of DNA Synthesis

*De Terra* [64] reported initiation of DNA synthesis in G1 nuclei of the ciliate *Stentor* after transfer into S phase cells, and the cell fusion studies of *Rao and Johnson* [188] and *Graves* [97] have shown that DNA synthesis can be induced in G1 nuclei in hybrids formed by fusion with S phase cells. Also, the fusion of G2 cells with G1 or S phase cells does not inhibit either initiation or continuation of DNA synthesis. These results, together with those of *Fournier and Pardee* [79], *Rao and Smith* [189] and *Rao and Sunkara* [190] imply that it is the production or activation of a critical initiating factor for DNA synthesis that is controlled at the transition point from G1 to S phase, and that once this initiator is present and active it allows DNA synthesis to proceed in a concentration-independent fashion. There is evidence from the data of *Cummins and Rusch* [60] on *Physarum* that the initiator molecules may be consumed in the process since limited DNA synthesis takes place if protein synthesis is blocked during S phase. It may be argued that the results obtained from experiments on organisms such as *Amoeba* and *Physarum* are not directly comparable to those relating to cultured cells since these organisms do not possess a G1 phase

[58, 59], but a similar phenomenon is found in the Chinese hamster V79 cell line [205]. In most cells the initiators of DNA synthesis are reported to be made towards the end of G1 [15, 185]. *Ley* [140] has drawn attention to a protein of $8 \times 10^4$ daltons molecular weight which is synthesised in early G1 and disappears at the commencement of S phase in growing CHO cells. Such behaviour would be consistent with an initiator of DNA synthesis which was used up in the process that it catalysed. Since the initiator appears to act in a fashion that is independent of the concentration it must be presumed that the process which takes place at the G1/S phase boundary is an activation of pre-existing initiator material. This could be brought about by the removal of inhibition, for example by the reduction to below a threshold level of the concentration of an inhibitor of initiation. The model of the regulation of cell size affords a mechanism sensitive to cell volume which is compatible with the evidence relating to the critical cell mass at the G1/S transition point discussed above, and which can be extended to include the evidence regarding the initiation of replication if the inhibitory substance were an inhibitor of the critical step in the initiation of DNA synthesis. To be compatible with the data from cell fusion studies the minimum hypothesis would appear to be that the inhibitor prevents either the synthesis or the action of the initiation factor. This might furnish an explanation for some of the contradictory results obtained in experiments using inhibitors of protein synthesis, since both the general effects on the rate of increase of cell mass and production of proteins essential for cell proliferation and the synthesis of the controlling initiator proteins would be affected by non-specific inhibitors of either transcription or translation. It may be, therefore, that the mechanism which the cell employs to regulate its size is one that operates through a threshold concentration of an inhibitor of initiator production for DNA synthesis. A threshold is implied by the existence of synchronization of DNA synthesis in cell hybrids. The 'sensing' mechanism might reside in a finely balanced titration of receptors with different affinities for the inhibitor.

An example of the type of molecular interaction envisaged are the different reactivities of various thiol species in the cell [234], and this is consistent with the argument regarding the probable influence of the relative concentrations of reducing substances on the origin of cell division. A good deal is known about the interaction of certain substances such as ketoaldehydes with thiol groups [217] but relatively little is known regarding the reactivity of polynucleotides.

### D. Response to External Factors

Irrespective of the precise molecular interactions that are involved, a model incorporating the features discussed above can be shown to possess the properties of constancy of cell size and the phenomenon of regulation in response to perturbations of the genome or cellular volume. The behaviour of the model is sensitive to a number of perameters, in particular the rate of increase in cell mass (which is regarded as equivalent to cell volume) which influences the length of G1.

It is known that the length of G1 is extremely sensitive to unfavourable growth conditions and arrest in G1 is the characteristic response of cells to medium depletion. Deprivation of essential nutrients such as leucine [72], isoleucine [139], glutamine [241], tryptophan [31] or valine [*Riley, unpublished*] results in cessation of growth of cells in G1. A critical mass model predicts that under conditions of nutrient deficiency G1 arrest will not occur at a particular stage and that on re-feeding the cultures the delay of entry into S phase will be dependent on the time taken to reach the critical size. Thus, rather poor synchrony would be anticipated after a block induced by amino acid depletion, as is found in practice, and in this way the process is analogous to attempts to induce synchrony in cultures by starvation [144] which were not very successful. The source of the variation is the cell volume at which protein synthesis is interrupted. It has been shown in mouse L cells that inhibition of protein synthesis by puromycin treatment for a fixed interval during early, middle and late G1 phase delays entry into S phase by a time equal to the duration of the inhibitory period [237] and the limited evidence available [134] suggests that cells maintain a constant division size at different growth rates.

## V. Extrinsic Factors

### A. Microenvironmental Effects (Intercellular Controls)
#### 1. Spatial Considerations

The most evident environmental limitation on the growth of a population of cells under favourable conditions of cultivation is the available space. In its most general form the limit is absolute in that there is a maximum number of cells of finite volume that can physically be accommodated in a given space.

The rate of change of a population that proliferates by binary fission

Control of Proliferation in Culture

in a time-homogenous manner can be represented by a function of the number of individuals that comprise it at any instant:

$$\frac{dN}{dt} = F(N), \tag{4}$$

where N is the number of cells in the population. Expanding the function as a Taylor series one obtains:

$$F(N) = pN + qN^2 + rN^3 + ..... \tag{4a}$$

For an upper limit of growth to exist, in keeping with the spatial argument above, the equation must have more than one root and the simplest case that satisfies this condition is:

$$F(N) = pN + qN^2 \tag{4b}$$

If the constant $p$ is interpreted as the sum of the intrinsic natality and death rates, i.e. the intrinsic net growth rate G, and $q$ as a negative constant equal to the ratio of the proliferation rate to the maximum size of population that can be contained in the available space then:

$$\frac{dN}{dt} = GN - GN^2/N_{max} \tag{4c}$$

which is the differential equation of the logistic:

$$\frac{1}{N}\frac{dN}{dt} = G(1-N/N_{max}) \tag{4d}$$

which states that the specific growth rate of the population is reduced by an amount which is proportional to the ratio of the population size at that instant and the maximum population size that can be accommodated in a given environmental location. While an intuitive understanding of the limits imposed on population growth by available space seems to have originated with *Linnaeus* [143] and, through *Malthus* [149], influenced *Darwin*, a general mathematical statement of environmental limitation was first made by *Verhulst* [248] in the form:

$$N_t = N_{max}/1 + \exp(u-vt) \tag{5}$$

and a similar equation was employed by *Pearl and Read* [177] to represent the growth of the population of the United States from 1790:

$$N_t = v . \exp(ut)/1 + w . \exp(ut). \tag{6}$$

*Winsor* [259] showed that the behaviour this equation approximates to the expression:

$$N_t = w . \exp(-\exp(u-vt)) \tag{7}$$

due to *Gompertz* [91] and there are many expressions which, more or less satisfactorily represent growth curves of cell populations [21]. The logistic equation has the advantage of simplicity and generates a sigmoid growth curve which matches data from many different types of culture with tolerable accuracy.

*a. Variation.* In any population there are variations in the behaviour of the individual that comprise it and variation in proliferation rates have been reported by many authors [7, 8, 199, 227, 245, 246]. Two divergent approaches to this variation exist. According to one, the variability exhibited by the population is the result of factors acting on a deterministic mechanism regulating the cell cycle in individual cells [132], whereas the alternative view ascribes the variation to the essentially probabilistic nature of the cell cycle [34, 226]. The original form of the *Smith and Martin* [226] transition probability model envisages the cell cycle is composed of an 'A' state and a 'B' phase. In the A state the cell does not progress to division but has a constant probability of transition into the B phase which is the conventional series of events in the cell cycle leading to mitosis in a constant period of time. The rate of division is therefore a function of the time taken for a cell to undergo the random transition from A → B. The evidence for this model is the exponential component in the frequency distribution of intermitotic times in growing populations of cells. A cell in the A state is regarded as being in an unstable condition analogous to a nucleus of a radioactive element which has a certain instantaneous probability of decay that is independent of the previously elapsed time. Thus, a population of cells in the A state will transit into the B phase and complete the next division with first-order kinetics and the time spent in the A state by any particular cell will be, by definition, unpredictable. According to the random transition model it is this random element which results in the heterogeneity of intermitotic times. The distribution of intermitotic intervals predicted by this model is an exponential fall in the fraction of cells ($\alpha$) having intermitotic times greater than a given time, after the time necessary to complete the sequence of events in B phase has elapsed. Allowing for some variation in the duration of B phase, this distribution fits published data from several sources tolerably well and elaborations of the random transition model to include probabilistic events in B phase [29, 30] can be expected to improve the correlation with the data.

The division products of a single progenitor cell would be expected to exhibit little variation in the time taken to complete the events in B phase and considerable attention has been given to sister-sister correlations in generation times since the distribution of differences of intermitotic intervals of siblings would be expected to reflect the differences in the time spent in the A state. The proportion of sibling pairs (β) whose intermitotic times differ by more than a certain value would, according to the transition probability model, be expected to correspond closely to an exponential distribution [163]. It is, however, possible to derive a distribution of sibling intermitotic time differences on the basis of a deterministic approach to the cell cycle using the critical size criterion for entry into S-phase and completion of the cycle. Assuming for any clone a constant final mass (M) then after cytokinesis each cell should ideally possess an initial mass of (M/2) but, allowing for variation about this value, the probability density function for the initial mass of any cell may reasonably be taken to approximate to a Gaussian distribution. Regarding the increase in cell mass as exponential, the time taken to reach the final mass (M) will be governed by the relation:

$$t_{(a)} = \lambda \cdot \log(M/m), \tag{8}$$

where $m$ is the initial cell mass and $\lambda$ is an inverse growth constant. For the sister cell:

$$t_{(b)} = \lambda \log(M/M-m). \tag{9}$$

Thus, the difference in intermitotic interval between siblings is a function of the inequality of mass at cell division:

$$t_{(a-b)} = \lambda (M-mm). \tag{10}$$

Substituting in the probability density function of the post-divisional mass ($m$) gives rise to β-plots of sister-sister differences in generation times of the form illustrated in figure 2a. This distribution corresponds more closely to

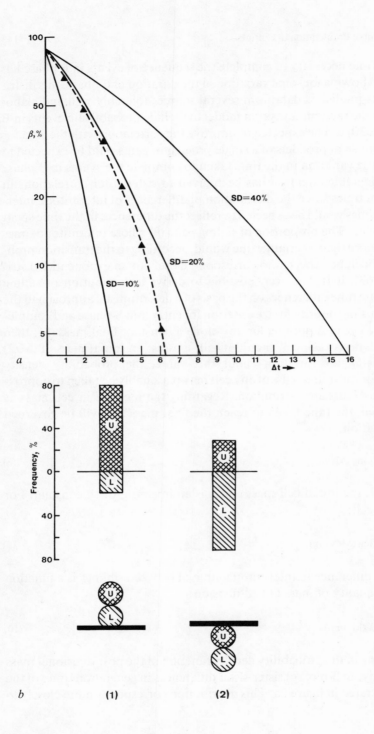

the observed sibling data than the random transition model [200, 245]. Indirect evidence to substantiate the idea that the post-divisional mass is a major determinant of the intermitotic time comes from experiments in which cells grown on an inverted surface were compared with cultures in the conventional orientation [201]. A proportion of cells divide in a plane vertical to the substratum and, if unequal division of cellular mass is posited, the influence of gravitation would be expected to introduce a bias in favour of the lowermost cell of a sibling pair of vertically dividing cells. Comparison of interdivision times of upper and lower cells arising from such division shows that this expectation is borne out; the lower cell in each orientation having a greater probability of having a larger post-divisional mass and therefore a shorter interdivision time than its sibling (fig. 2b). The importance of cellular mass in regulating cell cycle times has been recognised by advocates of the transition probability model. A source of variation not so far considered is the inconstancy of size at division [132] and it has been shown that cells deriving from small mother cells have longer cycle times than those from larger progenitors [219]. Some recent suggestions have been made which incorporate both deterministic and probalistic elements in the control of the cell cycle, such as the proposal by *Fantes and Nurse* [74] that a single structure is built by each cell in proportion to its size which then interacts with a receptor site, the interaction being described by a probability function of time. While this is an ingenious model there are formidable difficulties inherent in testing a scheme with an obligatory random stage on a statistical basis [130]. Other factors would also be expected to influence the distribution of generation times in a population of cells, in particular the accumulation of somatic mutations and other vertically transmitted characteristics which may account for much of the interclonal variation observed in cultures under

---

*Fig. 2. a* Semilogarithmic plot of the percentage ($\beta\%$) of siblings with differences of interdivision times greater than the time ($\Delta t$) shown on the abcissa. The curves shown for standard deviations in post-divisional mass of 10%, 20% and 40% about the mean were calculated as described in the text. The interrupted line corresponds to an SD of approximately 16% in post-mitotic mass and is fitted to the data points (▲) of *Van Wijk and Van de Poll* [245]. *b* Effect of relative position on the intermitotic interval of vertically mitosing GPK cells [from ref. 201]. (1) Lower (L) cell of vertical mitosis attached to substratum. (2) Upper (U) cell attached to substratum. The histograms show the frequency with which one division product exhibited a longer subsequent intermitotic interval than its sister cell.

similar environmental conditions (fig. 3a), and the differences in the local environment of individual cells may also be expected to contribute to variability of intermitotic intervals.

Taken overall, the growth rate of a population of cells may reasonably be expected to show a normal distribution about a mean value for the particular population under specified growth conditions. The intermitotic intervals in GPK cultures have been shown to correspond to a reciprocal transformed Gaussian distribution (fig. 3b) [167, 199] and evidence has been presented from continuous labelling of chick embryonic retinal cells with BUdR in vivo [200] that the variation in the cell cycle can be represented statistically by an inverse Gaussian distribution [78] and, moreover, that the variability is largely derived from one part of the cycle.

In a stochastic treatment of a logistic population, first attempted by *Feller* [75], it was found that the main effect of introducing statistical fluctuations was to slow down the growth rate in comparison with the deterministic model but the effect is negligible except with very small populations when random extinction is possible. A more recent and elegant treatment is to be found in *Bartlett* [14]. The general model proposed by *Bartlett* [14] is based on a probability-generating function $\theta_t$ (z) which for a time-homogenous system has the solution $\theta_t(z) = \theta_0(z) \exp(Ht)$ with a possible equilibrium condition: $H\theta(z) = 0$ where H is an operator acting on the probability generating function. The deterministic analogue is obtained by postulating a stochastic change during an interval $\triangle$ t such that:

$$\triangle N_c = N_{t+\triangle t} - N_t, \qquad (11)$$

with a probability given by R $(N_{t+\triangle t}/N_c) \triangle$ t. This is formally equivalent to:

$$dN = F(N)dt + dZ_t, \qquad (12)$$

for which the deterministic approximation $dN/dt = F(N)$ neglects the term $dZ_t$. Obviously the significance of this omission depends on the magnitude of the term. *Bartlett* [14, p. 14, equation 14] gives a calculation of the extinction probability.

An interpretation of the logistic growth equation is that it demonstrates the effect of the degree of saturation of a habitat where the upper limit is fixed so that there are increasing penalties against rising density of a population. Several different factors may be implicated, including inhibition of proliferation or increase in cellular death rate which would alter the net growth of the population. In most established cell cultures the death rate is small and does not vary greatly at different densities whereas there is abundant evidence that the proliferation rate of cells is reduced when the culture density rises.

*b. Contact Area.* In the case of monolayer cultures in which the cells are adherent to a solid substratum the surface that is available for cells to flatten on will be reduced with increasing density. Assuming a homogenous distribution, the average degree of cell flattening will be proportional to the unoccupied surface area so that the area of the cell adjacent to the substratum ($A_c$) can be described by the function:

$$A_c = F(\Phi, N/N_{max}),\tag{13}$$

where $\Phi$ is a cell flattening term, the value of which will reflect the adhesiveness of the cells and their ability to spread out on the material of which the substratum is composed.

*c. Adhesion.* The adhesion of cells to various substrata has been investigated for many years and the details are still not well understood. In general, cells possess surface groups capable of forming adhesive interactions with certain materials. Although it was earlier considered that electrostatic interactions between ionised surface groups were of major significance [61, 253] it now seems likely that the main adhesive interactions are contingent on the formation of multiple hydrogen bonds, a process that is assisted by the cooperative effects of localised domains of adhesive glycoproteins at the cell surface. Such adhesive plaques [5] are rich in cell coat material [57, 232] and are associated intracellularly with actin filaments [108] and other cytoskeletal elements. There is evidence that these specialised adhesive domains are stabilised internally by the cytoskeleton [191] and externally by polyvalent cations such as calcium [63]. The importance of hydrogen bonding for cellular adhesion to solid substrates is evident from the nature of adhesive and non-adhesive surfaces, and the requirement for 'wettability' which necessitates the modification of the surface of polystyrene culture ware. The material adherent to this wettable surface is a class of high molecular weight glycoproteins collectively known as fibronectins which are secreted by cells (table II). There is evidence from studies on certain lectin-resistant mutant hamster fibroblasts [179] that cells with reduced levels of surface carbohydrate receptors synthesise and release fibronectin but are unable to retain it at their

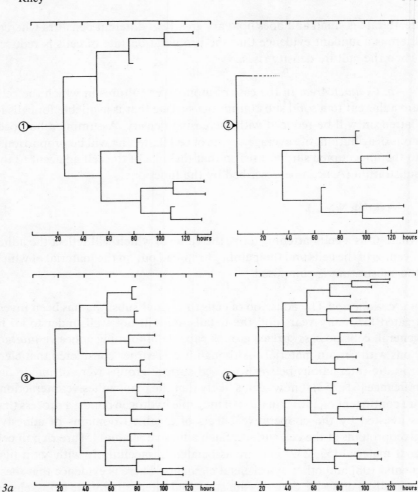

*Fig. 3. a* Dendrograms showing the interclonal and intraclonal variation in intermitotic interval in GPK cells cultivated in 10% serum-containing medium under subconfluent conditions [from ref. 199].

# Control of Proliferation in Culture

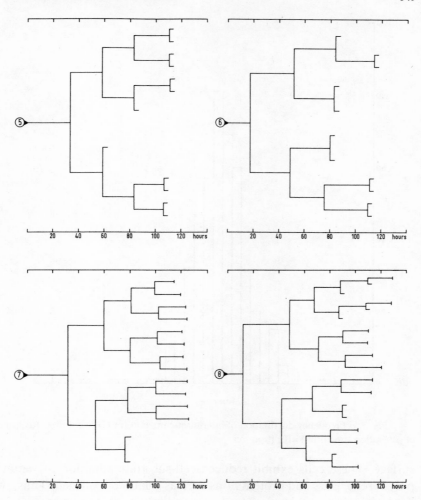

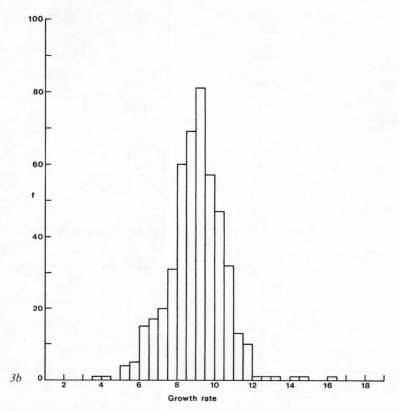

*Fig. 3. b* Frequency distribution of intermitotic intervals in GPK cultures. Reciprocal of generation time (min) ×10⁴ [from ref. 200].

surface. These cells exhibit reduced cell-substrate adhesion to wettable polystyrene, glass and protein films such as gelatin or collagen. Thus, two processes seem to be involved; the adhesion of fibronectin to the substratum, which seems to be a hydrogen-bonding interaction, and the interaction between fibronectin and cell surface receptors. In addition to carbohydrate binding other interactions may be of importance such as the formation of disulphide linkages or covalent linkages involving the carbohydrate residues of the interacting molecules although, since about half the surface fibronectin is removed by incubation of cells in 1 $M$ urea [263] and nearly all the surface fibronectin can be separated from plasma membranes by density gradient centrifugation [151], it is likely that most of the fibronectin is not covalently bound to the cell membrane receptors.

*Table II.* Fibronectin nomenclature [after 262]

| Glycoprotein factor | Reference |
| --- | --- |
| Cold-insoluble globulin | 168 |
| Anti-gelatin factor | 260 |
| Microfibrillar protein | 210 |
| Opsonic protein | 9 |
| Fibroblast surface antigen (FSA) | 215 |
| Galactoprotein a | 82 |
| Cell attachment factor (CAF) | 129 |
| Large external transformation-sensitive protein (LETS) | 121 |
| Cell surface protein (CSP) | 264 |
| Zeta (Z) | 206 |
| Cell-spreading factor | 101 |
| Fibronectin | 135 |

Most cells secrete fibronectin continuously into the culture medium but in sparse cultures the amounts available for cell attachment may be inadequate to coat the surface of the substratum and this may be responsible for part of the serum requirement of certain cells. This phenomenon may also furnish an explanation for the cooperative interactions between cells such as the 'contact spreading' described by *Middleton* [162] which could be accounted for by the increase in local concentrations of fibronectin in the immediate environment of a group of cells.

It is convenient to treat such short-range interactions as equivalent to collisional events (fig. 4). The collision frequency ($C$) of cells moving at random in two dimensions is a function of the number of cells: $C = f(N)$ where $f(N) = k(N^2 - N)$ which for large values of N is approximately equal to $kN^2$; $k$ being a constant which reflects the relative mobility of the cells on the available surface. Taking this into account in describing the spreading of cells on a surface, the function 13 becomes:

$$A^1_c = F(\Phi, k, N^2, N/N_{max}). \tag{14}$$

This function is discussed below in relation to the available metabolic exchange surface of cells in monolayer cultures. The significance of the terms $\Phi$ and $k$ require some elaboration, particularly as there are some zones of interaction between them. The mobility constant $k$ may be viewed

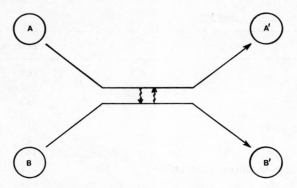

*Fig. 4.* Short-range cell-cell interaction viewed as equivalent to a collision. Cells A and B approach each other closely enough to enable exchange of 'information' indicated by the squiggly arrows.

as a kind of modified diffusion constant for cells [83, 84] and the mobility of cells is closely related to their adhesiveness to the substratum. Nonadherent materials, such as agarose films, immobilise cells while good adhesive surfaces permit rapid directional movement. Thus, a combined 'spreading' constant ($\Psi = \Phi k$) encompasses the influence of the substratum-fibronectin interaction [102], the amount of fibronectin available in the system, and the interaction between the cells and the fibronectin.

A major source of interest in fibronectin as the mediator of the adhesive interactions of cells has been the finding that the amount of cell surface-associated fibronectin is decreased in certain virus-transformed cell lines [262], although it is not clear in some cases whether this is due to a reduction in the amount of fibronectin elaborated by the cells or in the fibronectin-binding surface receptors. The mechanism of release of mitotic cells from a surface appears to involve a reduction in the numbers of fibronectin-binding sites and there is a good correlation between the reduced lectin agglutinability of mitotic cells [204] and a fall in cell surface-associated fibronectin at mitosis [121, 178]. Mouse 3T3 cells cultured in medium containing low concentrations of serum possess little fibronectin, possibly due to the small amounts added and the low density of cells. Treatment with epidermal growth factor (see later discussion) results in increased fibronectin production [47] and this may be an important factor in the mode of action of this growth-promoting hormone since other growth factors do not have this effect, with the possible exception of thrombin [169].

*d. Contact Inhibition of Movement.* In discussing the significance of factors influencing the adhesion of cells to substrata, and in particular in relation to the collisional events referred to above, it is important to recognise the influence exercised on the distribution of cells in a population by the characteristics of locomotory behaviour when cells come into contact with each other. Although the phenomenon of contact inhibition of locomotion had been known for many years, particularly in epithelial wound healing, it was first described in detail by *Abercrombie and Heaysman* [3], working with chick heart fibroblasts moving on a flat surface in tissue culture. They defined contact inhibition as the stopping of movement of a cell in the direction which had produced a collision with another cell, i.e. the prohibition, when contact between cells had occurred, of continued movement such as would carry one cell across the surface of another. This was shown to be a reaction in direct response to physical contact between cells. This reaction has been described in many different kinds of cells [6, 66, 161, 172]. The original data on contact inhibition of locomotion [3], were obtained from studies on the displacement of the whole cell. It was found that when explants were placed so that the cellular outgrowth met between them, further movement of cells in the direction which produced the collision was inhibited and cells tended not to overlap one another and as a result, a monolayer was produced. The formation of a more or less single layer of cells over the substrate and the virtual absence of nuclear overlap rapidly came to be used as criteria for the presence of contact inhibition. The degree of contact inhibition was expressed as an overlap index, which was a value derived by estimating the number of cell nuclei which would be expected to overlap if the cells were distributed at random and comparing this with the number of overlaps which actually occurred; the ratio of the observed number to the expected number being the overlap index. Populations of fibroblasts from different sources were shown to have significantly different overlap indices under standard culture conditions [6] and when dissociated cells of two heterologous fibroblast populations were cultured together, the presence of cells of one kind was shown to influence the homologous overlapping behaviour of the others. Several theories have been advanced to explain the phenomenon of contact inhibition [for reviews, see 2, 105]. Two main categories of theory have been advanced. One suggests that the phenomenon is a question of differential adhesiveness between cell and substrate and cell and cell. The second considers the phenomenon to be due to a direct inhibition of the cellular locomotory machinery which occurs upon collision. It may be that both these phenomena are involved. *Vesely and Weiss* [250] suggested the terms type 1 and type 2 inhibition; type 1 being typified by inhibition of the ruffling movements of the leading lamella and type 2, in which extensive overlapping does not occur, but in which no inhibition of ruffling activity takes place; *Di Pasquale and Bell* [66a] proposed the term 'inhibition of ruffling' for type 1 contact inhibition. It has been shown by *Abercrombie* et al. [5] that a cell does not need to ruffle in order to move. Cells moving under agar or methyl cellulose show no ruffles and the full implications of type 1 inhibition are not clear. Attempts to reproduce the phenomenon of type 1 contact inhibition when cells collide with non-living material have been unsuccessful. Although *Heaysman and Turin* [110] have used zinc-fixed chick heart fibroblasts to produce a typical contact reaction with living cells on collision, in view of the properties of polyvalent cations as cell adhesion factors noted above, it may be that the effect is not physiologically induced. Cells have been shown to be unable to move onto less adhesive substrates [40] but when confronted with boundaries of such substrates they have not been observed to undergo the inhibition of ruffling activity.

In contact inhibition type 2, the characteristic ruffling of the leading lamella is not inhibited and may exhibit greater activity on collision, suggesting the possibility that the cell is unable to make a satisfactory adhesion. *Carter* [40] suggested that if cell surfaces were less adhesive to one another than to the substratum material then the cells would tend not to overlap. Contact inhibition of locomotion would thus be a special case of haptotaxis. A similar view is taken by *Martz* et al. [152], who consider contact inhibition to be a special case of *Steinberg*'s [228] differential adhesion hypothesis.

*Abercrombie* et al. [4] reported that cells from mouse sarcomas, S37 and S180, did not exhibit contact inhibition when they came into collision with normal chick heart fibroblasts or mouse skeletal muscle fibroblasts, neither the sarcoma cells nor the fibroblasts being inhibited. It was at first assumed that this was a behavioural characteristic which showed a marked difference between normal and malignant cells. Although many subsequent authors have reported similar findings, it is likely, in view of the many indications of altered adhesiveness in transformed cells, that it is contact inhibition type 2 which is lacking in most of these cases. A point which requires stress is that it was not suggested by the original investigators that homologous contact inhibition would be lacking in tumour cells. The essential lesion that was identified was a reduction in heterologous contact inhibition. Most, if not all, tumour cells exhibit homologous contact inhibition and the importance of the relative degrees of homologous and heterologous contact inhibition have been emphasized [109].

From the point of view of a single population of cells in layer culture the existence of homologous contact inhibition of locomotion should ensure an approximately homogeneous distribution of cells in a culture. In those instances where contact inhibition is diminished the tendency for cells to heap up may result in an architecturally distinguishable uneven pattern and this will have some influence on phenomena, such as the available exchange surface, which are related to cell crowding.

*e. Cell Flattening.* Assuming a homogeneous distribution of a cell population on a two-dimensional surface and that the total available area is equivalent to $N_{max.}$ minimally flattened cells then the following considerations apply.

A minimally flattened cell may be regarded as columnar, cubical or (for tractability of the calculation) hemispherical, with a circular attachment area and a height equal to the cube root of $(3V/2\pi)$ where V is the cell volume. The cell height when a cell is fully flattened has been estimated and several reports [46, 77, 173] suggest that the minimum cell height of mammalian cells in layer culture is about 2–3 µm. Thus, for cells with a mean volume (V) the range of cell attachment areas can vary between approximately $A_{max} = (V - 2\pi/3)$ and $A_{min} = \pi(3V/2\pi)^{2/3}$, i.e. about 3-fold between 980 µm² and 360 µm² for cells with a mean cellular volume of 1,000 µm³. From the projected or contact areas it is possible to calculate the approximate exposed or exchange surface of the cell under different conditions of spatial restriction. This shows that the exchange surface decreases in a manner analogous to the contact surface and, thus, the effect of increasing cell numbers on the ratio of the cellular exchange surface to the volume is considerable between the limits set by the availability of space. If the uptake of essential nutrients by the cell is related to the available exchange surface and growth is proportional to it then the proliferation rate of a population of cultured cells can be described by the modified logistic equation in the form outlined above (fig. 5). This approach is entirely consistent with observations on the effect of substrate adhesiveness in cell proliferation [194] and the conclusions of *Folkman and Greenspan*

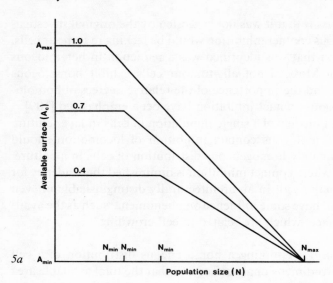

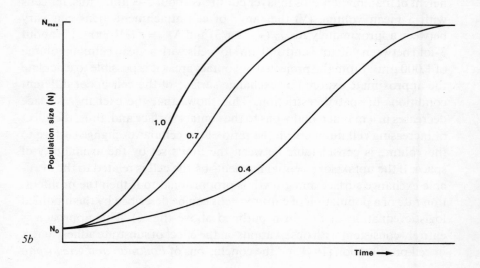

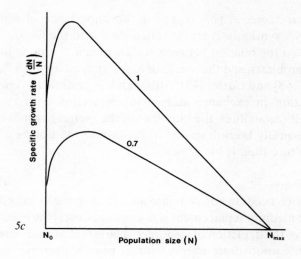

*Fig. 5. a* The form of the function ($A_c$) determining the available attachment area, and thus the exchange surface, of cells in monolayer culture. The effect of increasing $\Phi$ is indicated by the differing degrees of initial maximal attachment surface indicated by ($A_{max}$). The slope of the right-hand portion of the curve, which represents the relationship between cell attachment area and crowding is given by the logistic relation:

$$A_c = A_{max} \} 1 - [ (N-N_{min})/(N_{max}-N_{min}) \},$$

where $N_{min}$ = total surface area divided by $A_{max}$; and $N_{max}$ = total surface area divided by the minimum surface required by the cells ($A_{min}$). The relationship between the area in contact with the substratum and the exchange surface can be estimated by treating the flattened cells as lenticles (segments of a sphere) [173] or as cylinders of various degrees of flattening [37]. The latter treatment gives a direct relationship between exchange area and substratum contact area if the rim is excluded, which is a reasonable approximation under conditions of cell crowding. If the maximum height is limited to the diameter of the area of contact with the substratum then the minimum contact (and exchange) surface is given by the expression: $A_{min} = V/(V/2 \pi)^{1/3}$. Measurements on monolayer cultures have shown that the minimum cell height in fully flattened cells is about 2–3 μm for cells with mean volumes of approximately 1,000–1,500 μm³, i.e. between five and six times less than the maximum height. These somewhat simplified considerations establish a six-fold range of alteration in the cellular exchange surface in response to crowding as between the limits of $A_{max}$ and $A_{min}$. *b* Sigmoid growth curves derived from the function illustrated in *(a)* for $\Phi$ = 1.0, 0.7 and 0.4, respectively. *c* Specific growth rate as a function of population size if collision-induced spreading and space limitation are taken into account (curves for $\Phi$ = 1.0 and 0.7 shown). The degree of maximal spreading is indicated by the value of $\Phi$ shown on the curve.

[76] and *Folkman and Moscona* [77] regarding the importance of cell shape in regard to DNA synthesis. It is also concordant with the findings of *O'Neill* et al. [173] on the relation between surface area and growth in hamster and mouse fibroblasts, and the results of *Kolodny and Gross* [133], *Zetterberg and Auer* [268] and *Castor* [43]. Although it is possible to view the effect of a reduction in exchange surface in connection with the outward passage of cell metabolites, the influence of the exchange surface on cell growth has generally been discussed in relation to the uptake of nutrients from the culture fluid [114, 239].

## B. Resource Depletion

A major factor which needs to be taken into account, especially in cell culture systems where medium replacement is not continuous, is the effect on the growth of cells of the depletion of resources. The essential nutrients include oxygen and a carbohydrate energy source, usually dextrose, in addition to essential amino acids and vitamins. There is evidence that additional factors are also necessary for long-term survival, including fatty acids and trace elements and possibly other unidentified essential nutrients. The extent of the depletion of a resource depends on the rate of utilisation by the cells. Thus, for a population of N cells the rate of change in the amount of a resource is given by:

$$\frac{dL}{dt} = -\mu N, \qquad (15)$$

where L is the nutrient under consideration and $\mu$ represents its utilisation rate per cell. Hence,

$$L_t = L_o \exp(-\mu N t), \qquad (16)$$

where $L_o$ is the initial amount of the nutrient present in the medium.

If it is assumed that the availability of L is rate limiting for the proliferation of the population of cells the specific growth rate will be proportional to the expression on the right-hand side of equation 16. This yields a growth curve of the kind illustrated in figure 6. An important point in relation to this type of limitation on the growth of a population is that the resource under consideration is depleted irrespective of whether the cells are proliferating. This would be the case for oxygen and dextrose, for example. In the majority of culture media the amount of dextrose present is in excess and dextrose depletion is an unlikely cause of attenuation of growth in culture.

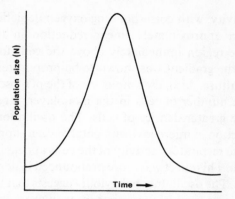

*Fig. 6.* Growth characteristics of a resource utilisation model in which a nutrient is used up continuously (see equation 16). The rate of reduction in the size of the population is a function of the death rate i.e. the magnitude of the negative component of the net proliferation rate.

### 1. Oxygen

Oxygen transport in cultures of attached cells depends on the kinetics of gas diffusion through the overlying medium. *Stevens* [229] calculated that the oxygen requirement and energy demands of a monolayer culture of liver parenchymal cells would not be met by the atmospheric partial pressure of oxygen ($pO_2$) if the depth of the fluid was greater than 0.34 mm. Calculations by *McLimans* et al. [148] showed that under standard medium overlay depth of 1.525 mm, oxygen depletion by a respiring monolayer of cells would exhaust the atmospheric oxygen supply at the plane of the sheet of attached cells in 35 min and this suggested that adaptive growth of cultured cells can occur only when respiration does not exceed the delivery rate of oxygen to the monolayer. *Werrlein and Glinos* [255] have investigated the changes in the medium $pO_2$ using oxygen-sensitive electrodes. Their studies were carried out on mouse fibroblast L cells grown in 50 mm diameter Petri dishes containing 6 ml of medium supplemented with 20% horse serum. The cultures were incubated at 35°C in a humidified atmosphere of 5% carbon dioxide and air. They detected oscillating concentration gradients of oxygen in the region just above the attached cell sheet and the oscillation amplitude increased with depth, i.e. as the measurements were made closer to the cells. It was shown by the use of an inhibitor of cellular respiration that the oscillations were an expression of cellular respiratory activity, with corresponding oxygen depletion in the microenvironment. An approximately linear reduction in the minimum $pO_2$ occurred in the region immediately above the cells to a height of 1 mm. The slope of the gradient was shown to be proportional to the numbers of cells in the culture. Also, the amplitude of the $pO_2$ oscillations were influenced by the number of cells in the monolayer, larger oscillations occurring with the greater density of cells. The oscillations, apparently, are due to the reduction in microenvironmental oxygen supply to low levels, at which point the respiratory activity of the cells appears to be affected by oxygen limitation. This effect was more pronounced in dense cultures than in sparse cultures. The oscillatory behaviour suggests that the entire cell population, or very large parts of it, respires in synchrony since individual cells respiring periodically but out of phase would cancel out the oscillatory effect. *Werrlein and Glinos* [255] therefore proposed that the cellular respiration in monolayer culture responds to environmental feedback and is modified by local exhaustion of essential nutrients such as oxygen.

## 2. Anabolic Nutrients

The kinetics for nutrients which are specifically utilised, or used at a much greater rate, during the increase of cell mass which is a prerequisite of proliferation are different. Thus, amino acids and precursors of phospholipids and nucleic acids, trace metals such as iron, and other factors are probably depleted from the medium at a rate approximating to:

$$\frac{dL}{dt} = -\mu \frac{dN}{dt} . \qquad (17)$$

Again assuming that the availability of a resource L is rate limiting for cell growth and proliferation, the resultant growth curve has the characteristics shown in figure 7.

By and large, the techniques that are used in cell cultivation are designed to obviate the effects of resource depletion of the type considered above. Experiments using different volumes of culture medium [46] suggest that, in general, the amounts of the medium components are large in comparison with the quantities taken up by the cells. In 3T3 cultures exposed to the same medium for 3 days and 8 days [37] only a slight shift in the distribution of the mean labelling indices was observed (fig. 8).

It is probable that if nutrient depletion is a significant factor in the control of growth and proliferation of cells it is most clearly expressed when there is reduced uptake by the cells [98–100]. Whether this is primarily the result of the reduction in exchange surface, as set out above, is open to question. Several early reports on the effects of increasing density on the rate of proliferation in cell cultures [42, 218, 266] indicate that the reduction in growth commences only at densities when most of the cells are in contact with each other, i.e. when some degree of spatial limitation becomes evident. The kinetics of this process are entirely consistent with the characteristics of the modified logistic outlined in equation 14 although it is not necessarily the case that the effect is exerted by a reduction in the exchange surface as is demonstrated for growth inhibition in normal mammalian melanocytes in vitro [195]. Phagocytosis is known to be cell-cycle depedent [198] and the importance of membrane movement and pinocytosis in nutrient uptake has been demonstrated [125]. The effects of contact inhibition, referred to above, would be to reduce such plasma membrane activity. Possibly a combination of factors is responsible for the alterations in nutrient uptake by crowded cultures [233].

Taking nutrient or growth factor depletion into account as well as spatial limitation it is possible to form an idea of the characteristics of the growth kinetics of cell cultures subjected to these, and other, influences and to compare them with the actual proliferative behaviour of cells (fig. 9).

### 3. Competition

Two important factors must be taken into account in considering the interaction between the individuals comprising a population subject to the environmental limitation. One is the range of the limit imposed; e.g. in the case of a spatial limit it obviously only applies to the individuals occupying a particular locality, and to exert a general influence movement of the individuals is implied and is analogous to diffusion. *Skellam* [224] has drawn attention to the limitations of diffusion models imposed by factors such as the texture of the medium, the pattern of movement, the relative ease and velocity of displacement and non-uniformity of movement but concludes that in general the analogy with heat conduction is useful in furnishing the notion of an equilibrium level which tends to be established everywhere in an enclosed system and local inequalities smoothed out. In the case of normal cells, i.e. cells subject to 'contact inhibition' of movement [3], growing on a restricted area the same mean density tends to become established throughout the system and thus the logistic equation limiting the size of the total population is a reasonable approximation to the effect of space limitation in monolayer cultures. Detailed studies of the changes in

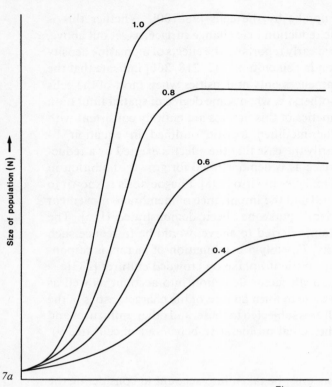

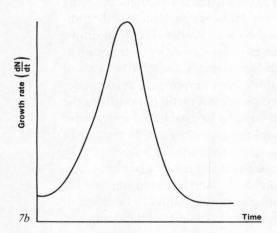

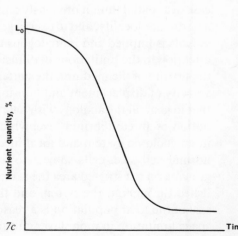

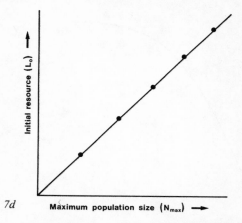

*Fig. 7.* Growth characteristics of resource depletion model (see equation 17), where the nutrient is consumed only during cell growth. *a* Growth curves for different initial resource levels ($L_o$ = 1, 0.8, 0.6 and 0.4). *b* Variation in growth rate with time. *c* Characteristics of resource depletion. *d* Linear relationship between the initial amount of resource ($L_0$) and the maximum population size (see also fig. 15).

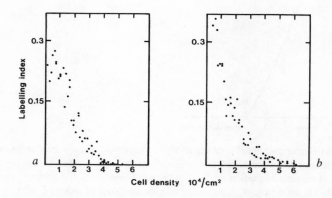

*Fig. 8. a* Scatter plot of the mean labelling indices observed at various cell densities in 8-day 3T3 cultures. *b* Scatter plot of the mean labelling indices in 3-day 3T3 cultures [data from ref. 37].

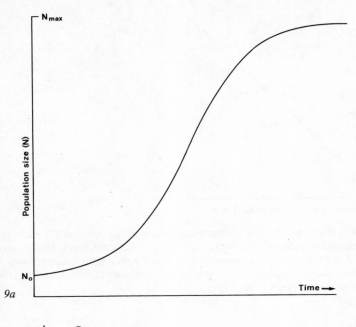

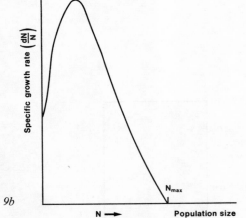

*Fig. 9. a* Growth curve calculated from the combined effects of cell crowding, collision-dependent cell spreading and resource depletion. *b* Specific growth rate as a function of population size. *c* Effect of resource limitation on $N_{max}$ under conditions of spatial limitation and collision-dependent cell spreading. *d* Comparison of characteristics of different models of growth limitation of monolayer cultures between the limits $N_1$ (= $N_{min}$; see fig.5a) and $N_{max}$. A = Spatial limit only; B = spatial + resource limitation; C = autoinhibition + resource limitation; D = spatial limit + autoinhibition; E = autoinhibition + spatial + resource limitation.

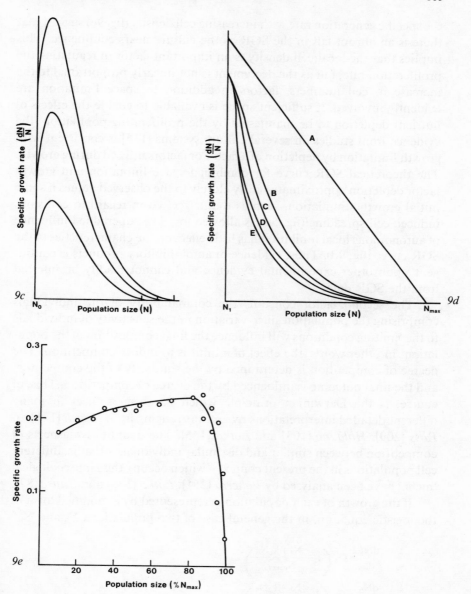

*e* Specific growth rate as a function of population size (% maximum population). The data are reciprocals of intermitotic intervals observed by time-lapse analysis of GPK cultures grown in polystyrene culture flasks. The segment for comparison with figure 9d is between 85% and 100% of the maximum population size.

the specific generation rate with increasing cell density (fig. 9e) suggest that there is an abrupt fall in the SGR as the culture nears confluence. This implies that the local cell density is an important factor in regulating the proliferation rate, but as the decrement is not linearly proportional to the increase in cell numbers, factors in addition to space limitation are evidently involved. If sufficient space is available to enable the effects of nutrient depletion to be manifested by the proliferating population, the evidence from studies on several culture systems [115] is consistent with growth limitation by depletion of a factor or factors utilised during growth. The theoretical SGR curve for combined space limitation and growth factor depletion approximates fairly closely to the observed values if some initial growth stimulation, such as that suggested in regard to collision-induced cell spreading (fig. 5c), is allowed for. The superadded influence of autonomous local inhibition has little effect on the characteristics of the SGR curve (fig. 9d). Thus, evidence of autoinhibitory products is contingent upon other experimental evidence and cannot readily be inferred from the SGR data.

The second factor concerns the comparability of the individuals comprising the population since variation in the sensitivity of individuals to the limiting conditions will influence the final composition of the population. In other words, the effect of a limit is to induce competition. The degree of competition is determined by the similarity of the competitors and the final outcome is influenced by the degree of competition. This, of course, is the Darwinian principle which has been subject to many different detailed interpretations by authors ranging from *Nägeli* [171], to *Ross* [209], *Haldane* [103] and *Lotka* [145]. The quantitative aspects of competition between similar and dissimilar individuals – that is, different cell populations in the present context – which occupy the same ecological 'niche' have been analysed by *Volterra* [251], *Lotka* [146] and *Gause* [87].

If the growth of each population is represented by a modified form of the logistic equation, in the general case of two populations $N_a$ and $N_b$:

$$\begin{aligned}\frac{1}{N_a}\frac{dN_a}{dt} &= G_a\left(1 - \frac{N_a + k_1 N_b}{N_{max\,(a)}}\right) \\ \frac{1}{N_b}\frac{dN_b}{dt} &= G_b\left(1 - \frac{N_b + k_2 N_a}{N_{max\,(b)}}\right)\end{aligned} \quad (18)$$

for which it can be shown that stable coexistence is possible only if $k_1 k_2 < 1$. The simultaneous differential equations state that each population reduces the available room for expansion of the other by an amount which is proportional to its own size, the coefficients of proportionality being indicated by the constants $k_1$, $k_2$ for $N_a$ and $N_b$, respectively. These equations define four equilibria. These are:

(i) $N_a = 0; N_b = 0$,
(ii) $N_a = 0; N_b = N_{max (b)}$,
(iii) $N_a = N_{max (a)}; N_b = 0$,
(iv) $N_a = \dfrac{(N_{max (a)} - k_1 N_{max (b)})}{(1 - k_1 k_2)}$; $N_b = \dfrac{(N_{max (b)} - k_2 N_{max (a)})}{(1 - k_1 k_2)}$.

If it is argued that for populations of cells the requirements for space and nutrients and other factors comprising the biotopic space will coincide rather closely so that: $k_1 k_2 \simeq 1$ then the conditions for stable coexistence do not exist. Values of k below unity indicate the extent of the zones of non-interaction. Provided these are small, the population size of the surviving, less successful, competitor will be negligible. The essential conclusion of this argument, the general validity of which has been confirmed by several authors [120, 155, 244], is that complete competitors cannot coexist. An important proviso for this to hold is that the distribution of competing species is homogeneous in the niche in order to ensure maximum interaction. In the case of cell populations this rarely obtains because of the clonal nature of propagation. Clearly the strongest competition will occur at short range, i.e. between directly adjacent populations, and this imposes limits on the *degree* of competition experienced and exerted in the system. Thus, two populations $N_a$ and $N_b$ arranged in blocks will compete only at the frontier between them with the result that even if $N_a$ is ascendant (i.e. is a more effective competitor), a substantial proportion of $N_b$ will survive. By contrast, if the populations are mixed then the proportion of $N_b$ surviving is diminished. It is possible to construct models of this type of system and show that the proportional survival of a relatively unsuccessful population is directly related to the degree of population mixing [*Vince and Riley,* in preparation]. Thus, by increasing the overlap the final proportion of the niche occupied by a population $N_b$ is reduced. Competition experiments of this kind using different cell lines have been made [*Riley,* unpublished] which confirm that survival of more slowly proliferating cells is proportional to the degree of mixing. This point may be of consequence to the question of the apparent lack of inter-clonal competition in tissues as evidenced by the data on G6PD mosaicism [35]. Although the abstract idea of a niche as a multi-dimensional space has

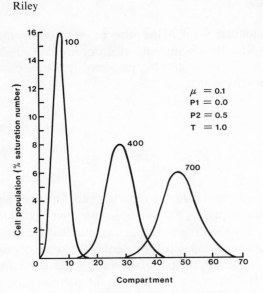

*Fig. 10.* Distribution of cells in compartments in conditions of zero loss from the population. The compartments represent decreasing levels of efficiency or 'fitness' of the cells and the constants refer to the mutation rate ($\mu$) the cycle time (T) and the probabilities of loss (P1) or damage (P2) [after ref. 196].

been thoroughly explored the central importance of actual spatial interference as a prerequisite to competition has not received much attention. The effect of spatial interference in forcing competitors into different niches has been identified but the extent of the limitation on direct competition, particularly in (monolayer) cell cultures deserves greater recognition. Spatial interference is to a considerable extent related to mobility. Thus, immobile organisms or cells will manifest less spatial interference than rapidly migrating ones, since less mixing is possible. The significance of social interactions such as contact inhibition of locomotion is broader than its direct action in promoting centrifugal spreading of cells because of the constraints that are placed on mixing of clones and thus on competition between them, since the only effective competition will be for unoccupied space at the margins. Thus, a growth advantage is not automatically translated into a selective advantage and depends for its expression on the availability of space and the relative location of the competing individuals. An interesting example of the limitation of competition has been found in insects by clonal analysis of the imaginal discs of *Drosophila* and *Onco-*

*peltus* [54]. There are certain boundary lines in the adult cuticle which are not crossed by clones after a given moment in development [32, 86]. Competition between such polyclonal 'compartments' is therefore no longer possible beyond certain stages in development. Whether a similar phenomenon occurs in mammalian development is not known. Another factor that requires to be considered in relation to competition between cells is the effect of the cessation of growth consequent on the population reaching its maximum size. Environmental limitation of the maximum size of a population can be shown to result in the curtailment of natural selection in a population [196] with the consequent accumulation of cells bearing heritable defects (fig. 10). Conversely, in cases where cell mixing occurs followed by a period of proliferation, as in the periodic subcultivation of cells in vitro, the effects of competition are likely to be manifested more rapidly [227] and may be responsible for the limit of growth potential (the 'Hayflick limit') displayed by cultures of diploid mammalian cells [107].

An interpretation which may be put on the condition $k_1 k_2 < 1$ is that stable homogeneous coexistence of two competing populations is possible only if at least one exerts a greater inhibitory effect on itself than on its competitor. Thus, to permit stability with non-zero values of $N_a$ and $N_b$ (for example, with steady-state populations of approximately equivalent size), when both populations are effectively occupying the same environmental niche, some form of specific self-limitation is necessary. In the case under consideration this can be expressed as:

$$\frac{1}{N_a} \frac{dN_a}{dt} = G_a \left(1 - \frac{h_1 N_a + N_b}{N_{max\,(a)}}\right)$$
$$\frac{1}{N_b} \frac{dN_b}{dt} = G_b \left(1 - \frac{h_2 N_b + N_a}{N_{max\,(b)}}\right) \quad (19)$$

where the condition for stable coexistence is $h_1 h_2 > 1$. The coefficient h may be regarded as composed of an environmental component (equal to unity) and an auto-inhibitory component (v): $h = (1 + v)$ so that for a population $N_a$:

$$\frac{1}{N_a} \frac{dN_a}{dt} = G_a \left(1 - \frac{N_a + v_a N_a + N_b}{N_{max\,(a)}}\right), \quad (19a)$$

where $v_a N_a$ is the inhibitory component specific to $N_a$, i.e. the *residual inhibition* exceeding any effect on the competing population $N_b$. Thus, $v_a = v_{aa} - v_{ab}$, where $v_{aa}$ is the effect of $N_a$ on $N_a$ and $v_{ab}$ is the effect of $N_a$ on $N_b$. Since the equation is concerned with inhibitory effects stimulation is denoted by negative quantities. It follows that if the stability condition is $v_a + v_b > 0$, then if one interaction is negative (i.e. residually autostimulatory) coexistence is conditional on a greater degree of autoinhibition in the competitor.

This method of expression separates the effect due to ecological competition from that due to an inhibitory effect which is specific to a particular sub-population. Thus, in the case of the two populations of cells $N_a$ and $N_b$ occupying the same fundamental niche the condition for stable coexistence is $v_a + v_b > 0$. A simplification of this approach is possible by writing:

$$a = G_a; \; b = G_a \frac{(1 + v_a)}{N_{max\,(a)}}; \; c = G_a/N_{max\,(a)},$$

and similarly:

$$d = G_b; \; e = G_b \frac{(1 + v_b)}{N_{max\,(b)}}; \; f = G_b/N_{max\,(b)}.$$

Thus, the specific proliferation rates become:

$$a - b\,N_a - c\,N_b$$
$$d - e\,N_b - f\,N_a$$

The stability of competing systems described by these expressions is described graphically in *Maynard-Smith* [156]. Fundamentally similar conclusions can also be derived from the general equations:

$$\left. \begin{array}{l} \dfrac{1}{N_a}\dfrac{dN_a}{dt} = F_{a1} - F_{a2}(N_a) - F_{a3}(N_b) \\[2ex] \dfrac{1}{N_b}\dfrac{dN_b}{dt} = F_{b1} - F_{b2}(N_b) - F_{b3}(N_a) \end{array} \right\} \quad (20)$$

At first sight this example of cellular altruism, i.e. voluntary self-containment, confers no advantage on the population which might otherwise dominate the niche and displace its competitor. It can be argued that by making coexistence possible the self-inhibitory property defends the population from the possibility of extinction by a more successful competitor. But the full significance of this property is probably to be found in its biological potential as a conditio sine qua non for the evolutionary step from single celled organisms to metazoa, because it is manifestly an indispensable prerequisite for cooperation between different cell types which are to share the same physical and ecological niche.

The genealogical evidence [7, 8, 199, 200] shows that clones differ in their intrinsic proliferative rates and thus competition between clones is a factor influencing the growth of a culture.

## C. Metabolic Products

Observations on the pattern of $^3$H-TdR incorporation in murine 3T3 cultures have demonstrated that density-dependent inhibition of proliferation is exhibited in relatively sparse cultures, commencing at about 0.5 x $10^4$ cells/cm$^2$ [37]. This evidence that a self-limiting process is in operation at densities where there is no spatial limitation or evidence of significant resource depletion suggests that there are metabolic products of cells which exert an inhibitory influence over the rate of growth of neighbouring cells. *Yeh and Fisher* [266] showed evidence for a heat-stable diffusible growth inhibitor produced by cells in culture. The material was butanol extractable and the molecular weight was estimated as less than $10^3$ daltons. A similar water-soluble material has been extracted from liver [158, 221, 249] which is precipitable with 70–87% ethanol and is sensitive to protease digestion, but most other growth inhibitory materials extracted from cells, with the possible exception of the erythroid [18] and granulocyte inhibitor [147, 175], do not exhibit the same properties [12]. It is probable that the technical difficulties inherent in the extraction and isolation of peptides from cells has tended to cloud the issue since, on homogenisation, lysosomal proteases may act on cell proteins to give rise to many types of peptide fragment which may exhibit growth inhibitory properties. A good case for the concept of tissue-specific inhibitors (chalones) exists and experimental evidence to support their regulatory role in vivo has been advanced [136] but their action has not been unequivocally demonstrated in cell cultures.

Metabolic products released into the culture medium may include peptides with relatively specific actions on certain cells but it is sufficient, in the context of autoinhibitory influences acting on a local sub-population of cells [193], to propose that inhibitory effects are exerted by non-specific agents such as, for example, urea, lactate or carbon dioxide. Exper-

*Table III.* Long range inhibition. Relationship between the mean labelling indices at different local densities and the estimated overall cell density on a series of slides inoculated heterogenously with 3T3 cells

| Mean overall density (× 10⁴ cells/cm²) | Local cell density range | | | |
|---|---|---|---|---|
| | A | B | C | D |
| 0.6 (31) | 2,680 (13) | 1,676 (5) | – | – |
| 0.9 (40) | 2,944 (16) | 2,918 (13) | – | – |
| 1.7 (33) | 2,257 (5) | 1,843 (22) | 186 (2) | 85 (1) |
| 1.9 (25) | 3,400 (3) | 2,057 (5) | 202 (2) | – |
| 2.0 (19) | 946 (2) | 599 (9) | 16 (2) | – |
| 2.1 (28) | 3,223 (1) | 2,967 (14) | 62 (2) | – |
| 2.2 (24) | 498 (3) | 592 (12) | 19 (2) | 7 (2) |
| 2.6 (19) | 1,522 (2) | 1,060 (5) | 10 (6) | 7 (1) |
| 2.8 (21) | 1,325 (1) | 1,571 (4) | 164 (7) | 132 (3) |
| 2.8 (25) | 1,526 (3) | 1,426 (4) | 26 (7) | 10 (2) |
| 2.9 (18) | 1,307 (2) | 271 (3) | 10 (3) | 7 (3) |
| Unweighted regression values: | | | | |
| Corr. coeff. (r) | 0.68 | 0.47 | 0.44 | 0.19 |
| Slope | –981 | –515 | –82 | –22 |
| Intercept (x = 0) | 4,064 | 2,643 | 268 | 97 |

The numbers in parentheses indicate the number of slides from which mean values were obtained. Four ranges of local cell density are shown marked in columns A, B, C, D which encompass local cell densities of 0–1, 1–2, 2–3, and 3 × 10⁴ cells/cm², respectively. The labelling indices are shown throughout as whole numbers by multiplying the actual labelling index by a factor of 10⁴. The calculated slopes are shown in figure 11 by a *heavy line* for set A, a *heavy dotted line* for set B, a *light continuous line* for set C, and a *light dotted line* for set D. The coefficients of correlation (r) are also indicated in Figure 11 which illustrates the considerable scatter of the data and the paucity of low density values in sets C and D [from ref. 38].

iments on non-homogeneously seeded monolayer cultures of 3T3 cells provide evidence that diffusible factors reduce the proliferation rate of cells in localised zones of the culture [38]. Some of these results (table III) demonstrate the existence of both strong short-range and weak long-range intercellular effects in layer culture, the latter influence being demonstrated by the effect of the overall population density in the culture on the cellular response at different local densities (fig. 11). It is probably not too

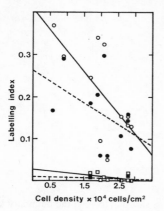

*Fig. 11.* The mean local labelling index at different densities as a linear function of the average overall density (data from table III). Sets of data: ○ = A; ● = B; □ = C; ■ = D.

great an oversimplification in the case of medium in a culture vessel incubated at 37 °C to regard the mixing in the fluid compartment as being brought about by convection and to be fairly uniform, while the boundary layer is regarded as a stagnant compartment [182]. The approximate thickness of the boundary layer under normal culture conditions has been estimated as $12.33 \pm 2.73 \mu m$ [38].

The distribution within the boundary layer of a diffusible factor released from a cell at a constant rate can be calculated from the analogous situation of heat flux in a flat plate with unilateral loss. The diffusion in a plate of constant cross-sectional area is given by:

$$\frac{\delta^2 C}{\delta x^2} - m^2 C = 0 \tag{21}$$

which has the solution for the concentration ($C_x$) at a distance (x) from the source:

$$C_x = K_1 \exp(-mx) + K_2 \exp(mx) \tag{22}$$

where $K_1$ and $K_2$ are constants and the parameter m represents a partition coefficient describing the ratio of the loss from the upper surface of the compartment and the flux through the layer and is defined $m = \sqrt{h/ky}$, where h = the mass transfer coefficient which is the rate of flux per unit area

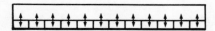

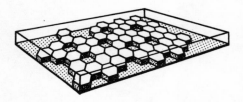

*Fig. 12.* Idealized division of fluid compartments showing (below) the boundary layer represented as a set of interlocking hexagonal pericellular compartments which intercommunicate (above) via the bulk phase of the culture fluid (supracellular compartment) thus permitting long-range chemical interactions to take place between cells in different areas of the culture system [after ref. 38].

per unit of concentration gradient between the boundary layer and the upper compartment, k = the diffusion constant of the boundary layer, and y is the depth of the layer. Alterations in the concentration gradient between the fluid compartments will be reflected by a nearly linear change in the absolute concentration in the boundary layer, which is consistent with the finding shown in figure 11. The gradient profile within the boundary layer is not influenced by this, as is indicated by the normalised equation if losses from the edge of the layer are neglected. This is:

$$C_x = C_0 \} \cosh m (L-x)/\cosh mL \tag{23}$$

where $C_0$ is the concentration at the source (where $x = 0$) and L is the normalised length of the compartment.

In the case of a metabolite which acts as an inhibitor of proliferation diffusing in the boundary layer, the relationship between the proliferation inhibitory effect and the distance from the source of inhibition is approximately linear over a short range. The relationship between the inverse square root of the cell density and the degree of inhibition of labelling index from data previously published on density-dependent growth inhibition in 3T3 cells [37] has been shown to be linear up to an intercellular distance of about 140 μm [104]. It is possible, on the basis of these measurements,

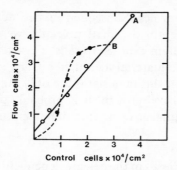

*Fig. 13.* The relationship between the final cell density of static and recirculated cultures of 3T3 cells. Each point represents the paired values of one experiment (recirculated cell numbers on ordinate and static cell numbers on abscissa). Results of experiments with serum-supplemented medium are shown as open circles (curve A) and those of serum-free experiments as filled circles (curve B). The seeding density was $5 \times 10^3$ cells/cm$^2$ in all cases. The sigmoid curve (dotted) was theoretically derived from the exchange model of *Canagaratna* et al. [38] assuming that the growth rate is inversely proportional to the concentration of an inhibitory factor generated by the cells at a constant rate, and scaled to correspond to the cell densities determined by experiment [data from ref. 46].

to envisage the boundary layer compartment as represented by an array of hexagonal pericellular domains (fig. 12), each of which communicates via a general supracellular compartment. This model confers on cells in a particular pericellular domain considerable autonomy of growth, but they are influenced from outside by the effect of concentration gradient on the diffusion rates in the fashion outlined above. This bicompartmental model indicates the way in which long-range influences of the overall cell population may be exerted in a way that we may regard as a biological version of Mach's principle. The importance of the boundary layer of medium in monolayer cultures has been demonstrated by the studies of *Rubin and Rein* [213], *Stoker* [230], and *Stoker and Piggott* [231].

In studies of long-range density effects on the proliferation rates of 3T3 and RLCW cells in recirculated medium [46], it was shown that in serum-free medium the effect of perfusion of 3T3 cultures was confined to a small density range and was consistent with the more rapid removal of a diffusible inhibitor from the pericellular environment by recirculating the medium (fig. 13). Treatment of the perfusing medium with immobilised proteases failed to produce convincing evidence to suggest that the inhibitory factor released by 3T3 cells into the circulating medium was susceptible to attack by protein-degrading enzymes. Subsequent experiments [*Riley* et al., unpublished results] have identified a small molecular weight material which is present in culture medium exposed to cells. This factor is heat stable and may correspond to the material identified by *Yeh and Fisher* [266]. It is not clear, however, that this material is the only inhibitory factor with a significant action on cell growth. It is probable that several metabolites are implicated in mutual growth inhibition in culture.

### 1. Hydrogen Ion Concentration

Many of the essential metabolic processes in the cell are dependent on maintaining a hydrogen ion concentration near neutrality. Ion transport systems enable cells to control their internal pH by exchange of $H^+$ with $Na^+$ or $K^+$ and, therefore, the extracellular pH is important in influencing the behaviour of cells. *Eagle* [69, 70] performed experiments to investigate the effect of extracellular pH on the rate of protein synthesis and plating efficiency of normal and virus-transformed cell lines. Normal human cell lines exhibit a sharp pH optimum at an external pH of about 7.7. Small variations ($\pm 0.2$ pH units), on either side of this optimum were shown to reduce plating efficiency by about 20%. To what extent this effect is due to the influence of pH on ionisable groups on the cell surface which may be of importance in altering receptor affinities is not certain. *Eagle* [69, 70] found that the effect of serum was influenced by the pH of the culture medium and this may be due to alterations in binding of growth stimulatory factors (see below). Similar results on the effect of pH on the growth-stimulating action of serum were reported by *Froehlich and Anastassiades* [81]. The pH supporting the highest growth rate was shown to be independent of the concentration of serum used in the experiments. Cell lines from other mammals show a similar response to the external hydrogen ion concentration. That some of the action of $H^+$ is external is shown by data that the intracellular pH of human cells is near 6.9 [150]. *Ceccarini and Eagle* [44, 45] studied the effect of extracellular pH on density-dependent growth inhibition in human diploid fibroblasts. At the pH which was found to be optimal for cell growth, maximum cell densities ($N_{max}$) were between 2 and 4 times as high as those present in less well-buffered systems. With bicarbonate buffering maximum cell densities of up to $1.5 \times 10^5$ cells/cm$^2$ were obtained compared with medium buffered with non-volatile, high-capacity buffers which supported growth to a cell density of up to $5 \times 10^5$ cells/cm$^2$. The stability of the pH of the culture medium was found to be a major factor in determining the saturation density of the system. The suggestion that local inhibition of proliferation may result from the excretion by the cells of acidic metabolites has also received support from experiments reported by *Lie* et al. [142]. *Taylor* [235] and subsequently *Rubin* [211] demonstrated that the effect of $H^+$ greatly exceeds the effect of carbon dioxide or bicarbonate. Cells exposed to suboptimal pH are more susceptible to inhibitory effects of local cell density [44]. Some of those effects may be mediated by modification of the adhesiveness of the cells or change in nutrient uptake consequent on surface changes.

Virus-transformed cells and spontaneous tumours in cultivation appear to show a less pronounced external buffering requirement and generally have a diffuse and rather lower pH optimum (near 7.2) than normal cultures [44, 45, 70, 211]. *Gerson* [88] has reported experiments in which cells treated with acetazolamide to inhibit intracellular carbonic anhydrase, and thus reduce the generation of $H^+$ from carbon dioxide and water, exhibited a lower external pH optimum for cell growth. This suggests that intracellular $H^+$ homeostasis is dependent on extracellular buffering.

If a close relationship exists between the internal concentration of $H^+$ and the pH of the surrounding microenvironment of the cell it is possible that localised autoregulatory interactions of the type outlined in the section above could be modified through local pH changes. No satisfactory data exist regarding the possible regulatory significance of changes in intracellular pH in tissue culture cells but *Gerson and Burton* [89] have presented evidence from intracellular pH measurements on *Physarum* that there is a cyclical variation in the hydrogen ion concentration. Their data (fig. 14) indicate that the mean pH is rather lower than that reported for human cells and fluctuates through the growth cycle betwen 5.8 and 6.4 reaching the highest pH at mitosis. Since in *Physarum* the S phase begins directly after division it is possible that the raised pH is a factor in the initiation of DNA synthesis and might, in mammalian cells, be of significance in determining the length of G1. A number of experiments favour this view. *Siskin and Kinosita* [223] showed that the G1 phase of the cell cycle was the phase most influenced by external pH, and *Rubin* [211, 212] demonstrated that both initiation and the rate of DNA synthesis was increased in cells exposed to medium with a higher pH. Also, *Calothy* et al. [36] have reported that the production of SV40 in monkey-mouse hybrid cells is pH dependent. *Johnson* et al. [123] have shown that the intracellular pH of sea urchin eggs rises as a result of a $H^+/Na^+$ exchange within a few minutes of fertilisation. It is, therefore, conceivable that loca-

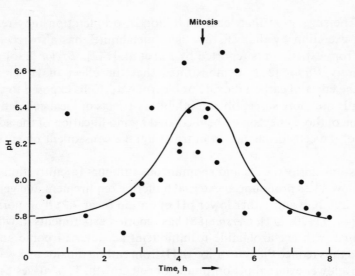

*Fig. 14.* Changes in intracellular pH in relation to the cell cycle in *Physarum* [data from ref. 89].

lised self-inhibition could be exerted by alterations in microenvironmental pH which, by affecting intracellular hydrogen ion concentration, may alter the rate at which reactions with regulatory significance occur.

## VI. Interpopulation Controls

### A. Hormones

*Todaro* et al. [242] showed that the addition of serum to a monolayer of resting 3T3 cells led to the initiation of DNA synthesis in some of the cells. *Holley and Kiernan* [115] demonstrated that the maximum size of a population of cells was determined by the amount of serum that was added to the culture. This was the case both for normal cells and for SV40-transformed 3T3 cultures, although the latter manifested a greatly reduced dependence on serum factors (fig. 15). Much of the subsequent progress in this field has resulted from work on fractionation of serum and the isolation and identification of particular factors with growth-stimulating

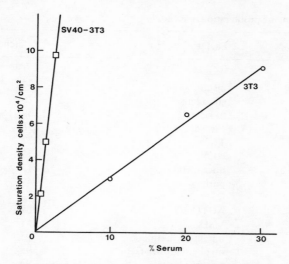

*Fig. 15.* Relationship between initial serum concentration and saturation density of 3T3 and SV40 3T3 cells [data from ref. 115].

properties. The results have demonstrated a plethora of hormones and growth factors.

It may be helpful to distinguish between hormones with a general action, such as insulin, and more or less specific growth-stimulating factors. In table IV the agencies listed under the first category may be regarded as forming some of the factors necessary for the growth and survival of mammalian cells which, by virtue of their evolutionary development, have become reliant on the specialised products of each other. This is naturally related to their degree of differentiation and may vary for different types of cells. The agencies named are usually required for cell growth and thus form the hormonal 'background' in which cells are able to proliferate. Adequate concentrations of these factors are normally present in the serum employed in the preparation of tissue culture medium but they may require to be added as supplements to medium containing deficient or low levels of serum. Thus, glucocorticoids [238], insulin and iodothyronines [137], linoleic acid [113], zinc [212] and purines [49] are able to restore growth under conditions of serum deficiency. One of the

*Table IV.* Hormones influencing proliferation in cells

| Classification | Factor | Reference |
| --- | --- | --- |
| *General hormones* | | |
| a Protein | insulin | 137 |
| | glucagon | 220 |
| | transferrin | 159 |
| b Steroid | cortisol | 238 |
| c Other factors | $T_3, T_4$ | 137 |
| | UFA | 113 |
| | trace metals | 212 |
| | purines | 49 |
| *Target-specific hormones* | | |
| a Protein | ACTH | 187 |
| | MSH | 261 |
| | luteinising hormone | 93 |
| | myoblast growth factor | 95 |
| | erythropoietin | 92 |
| | CSF | 160 |
| | mesenchymal factor | 138 |
| | prolactin | 138 |
| b Steroid | oestrogen | 26 |
| | progesterone | 202 |
| *Growth factors* | | |
| | epidermal growth factor (EGF) | 52 |
| | fibroblast growth factor (FGF) | 11 |
| | platelet growth factor(s) | |
| | somatomedin C | 247 |
| | CRL-MSA | 225 |
| | Cationic serum polypeptide | 10 |
| | ovarian growth factor | 48 |
| | prostaglandins | 122 |

major characteristics of transformed cells is their reduced requirement for serum and this is also true of some individual factors.

In the case of cultures derived from tissues which in vivo depend on target-specific hormones, such as mammary tissue, it is often necessary to add the appropriate hormones to the culture medium. In the case of the breast it is the fibroblasts which respond to hormone treatment and not the epithelium [202, 203].

Superimposed on this complex background of essential nutrients, hormones, co-factors, vitamins, etc., are the growth-stimulating effects of

*Table V.* Proliferation-controlling hormone receptors

| Receptor | Reference |
| --- | --- |
| Insulin | 56 |
| Glucagon | 208 |
| Epidermal growth factor | 112 |
| Somatomedin | 111 |

a range of factors, many of which form a category of polypeptides with a molecular weight range of 5–15 kilodaltons. It has been suggested that they are related by partial primary sequence homology [111]. In addition there are smaller fragments, such as the tripeptide (Gly-Lys-His) reported to stimulate proliferation in hepatoma cells in culture [180] and the low molecular weight brain factors of *Koch* et al. [131]. The burgeoning scene of growth factors implies a very wide range of these substances.

At present it is not clear to what extent these growth-stimulatory factors act independently of each other. It is known that many cells exhibit EGF receptors [112] and the biological effect of EGF is proportional to the degree of binding to the plasma membrane receptors. Separate plasma membrane receptors have been identified for a number of the hormones listed in table IV and these are summarised in table V. In addition, cytoplasmic and nuclear oestrogen-binding receptors and cytoplasmic glucocorticoid-binding proteins [27, 243] exist in some, perhaps all, cells.

Evidence that cultures are able to grow in the absence of serum if the medium is supplemented with appropriate growth factors comes from work with the human mammary cell line MCF7. The rate of proliferation of cultures grown in serum-free medium containing physiological levels of insulin, epidermal growth factor, transferrin, prostaglandin $F_2$ and fibronectin was identical to that of cultures in medium supplemented with foetal bovine serum [13]. Also, *Honegger* et al. [116] have shown that foetal rat brain cells will incorporate $^3$H-TdR into DNA in a serum-free medium supplemented with insulin, cortisol, triiodothyronine, transferrin and trace elements but their growth was less rapid than that of similar cultures in serum-supplemented medium.

Not all hormonal factors that have been identified are growth stimulators. For example, ACTH has an inhibitory action on the target cells of the adrenal cortex [187] and glucocorticoids are inhibitory to fibroblast growth [141]. Moreover, there are a number of inhibitory compounds with varying degree of tissue specificity most of which are heat labile and degraded by proteases [12]. Although specific receptors for these agents have not yet been identified it is probable that they exert their growth-inhibitory effect by binding to sites on the plasma membrane.

## 1. Mode of Action

Most of the hormonal agents discussed above exert their action in modifying the metabolism of the cell by suface receptor binding. Even the growth-stimulating effect of calcium [68] appears to be an indirect action, possibly by altering the receptor binding of serum-derived growth stimulators. Some evidence for this comes from the calcium-sensitive period (CaSP) which occurs between 6 and 8 h after serum stimulation of quiescent cultures when calcium removal by transient washing with 2.5 m$M$ EGTA leads to the loss of a serum component which is necessary for the continued growth of the culture [63].

Some degree of understanding of the hormonal effects on cells has come about through knowledge of the changes in cyclic nucleotide levels in response to receptor binding. Certain peptide hormones with an inhibitory activity such as ACTH [187] and gonadotrophin cause rises in the intracellular concentration of cAMP. Gonadotrophin addition to Chinese hamster ovary cells in culture showed a maximum increase of intracellular cAMP levels 30 min following exposure to either human chorionic gonadotrophin or follicle stimulatory hormone, and returned to basal levels in 2 h. The enhancement of cAMP concentration was dependent on the hormone concentration in the medium and slowed the proliferation rate of the culture accordingly [71]. There is general agreement that raised intracellular levels of cyclic AMP inhibit proliferation and a direct relation between the cellular cAMP levels and generation (doubling) time has been observed [24, 153, 240, 257] although the correlation is not consistent in all systems [176].

Some stimulatory receptor-binding hormones stimulate the production of cGMP [1, 94, 214]. Prostaglandin $F_{2\alpha}$ has a growth-stimulating action on confluent 3T3 cells which is correlated with a small increase in cGMP and a concurrent reduction in cAMP [122]. The greater sensitivity to change which is exhibited by a ratio of cyclic nucleotides has led to the 'Yin-Yang' hypothesis [90] that it is the ratio of the levels of cGMP and cAMP that exerts the controlling function. The mechanisms involved in this control are far from clear. Work with mouse S49 mutant cells [51] suggests that the point of action of cAMP is a cAMP-dependent kinase.

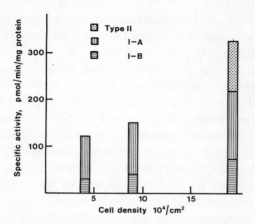

*Fig. 16.* Mean phosphodiesterase activities in RLC cultures at varying cell density. Enzyme type IA = sulphydryl-independent, theophylline resistant activity; type IB = sulphydryl-independent, theophylline-sensitive; type II = sulphydryl-dependent activity. Assays were made under four conditions: (1) in the presence of dithiothreitol $10^{-4}M$; (2) in DTT plus theophylline $10^{-3}M$; (3) in DTT plus N-ethylmalleimide $10^{-3}M$; (4) in DTT, theophylline and NEM [after ref. 197].

Whether this action is influenced by the cGMP levels, e.g. by competition for the cyclic nucleotide receptor site, or whether cGMP has other actions such as increasing the rate of cAMP hydrolysis [166] is not known. Data on rat liver cell cultures [197] have indicated a rise in intracellular phosphodiesterase levels in response to increases in cell density (fig. 16). The biphasic pattern of density-dependent inhibition of proliferation in RLCW cells may be due to activation of a receptor-activated phosphodiesterase. In the case of lymphocytes the effect of growth stimulation by surface-binding lectins is not associated with changes in the levels of cyclic nucleotides [252] and the mechansim of stimulation may involve changes in intracellular calcium concentration [256]. The relationship between intracellular cation concentrations and cyclic nucleotides may be complex. An attractive scheme for the action of hormones has recently been proposed by *Rodbell* [207].

*Moens* et al. [166] reported that there was no rise in cAMP levels in fibroblast cultures as they become confluent and other evidence [117, 118] implies that the action of hormones and other modifying agents on the proliferative behaviour of cells is superimposed on the basic controls. The strongest evidence for this comes from the single-step cloned mutant lymphoma lines deficient in cAMP-dependent protein kinase since these cells maintain a normal cell cycle [25, 51].

There are many changes that take place in cells during transition from quiescent to the growing state in response to stimulation. It seems likely that these steps are normally controlled intrinsically and modified by external factors. In terms of the hierarchical system of regulation proposed here the action of the microenvironment may be viewed as a combination of restraints on space and essential nutrients and autoinhibitory effects exerted by local accumulation of metabolites which diminish the efficiency of the anabolic processes necessary for continued growth. In *Stoker's* apt phrase the cells seem to be able to 'smell each other'. The superimposed action of hormones can be interpreted in the light of modifications of any of the sizeable array of metabolic events involved in the increase in cell mass necessary to trigger entry into S phase and the progress to mitosis. Of these, perhaps the most influential in re-initiating cell growth would be activation of membrane-transporting systems [90, 112, 122, 137, 214] either by a direct effect which would by rapidly exerted or by the stimulation of their synthesis or turnover.

### 2. Transformed Cells

Evidently, cells with a constitutively greater uptake of nutrients will tend to grow to higher population densities and this is one of the consistent characteristics of transformed cells in culture and such cells also exhibit a reduced sensitivity to available exchange surface [173], lower sensitivity to pH changes [44, 45, 70, 211], and reduced requirements of stimulatory factors [115]. While it is important to distinguish between 'transformed' and spontaneously arising tumour cells, diminished sensitivity to environmental restriction would furnish neoplastic cells with a growth advantage over normal cells. Further studies with temperature-sensitive mutant strains may help to unravel the complexities of the control of proliferation [181]. In view of the several actions attributed to EGF it would appear improbable that the resulting picture will be simple.

## *VII. Conclusions*

The control of proliferation in cell populations has been analysed on the basis of a hierarchical system of influences which modify a fundamental intracellular switching system which adjusts the timing of cell division to coincide with a doubling of the genome and the cell mass. Evidence to show that the rate of increase of cell mass is a basic limiting factor is discussed, and the nature of the permissive extrinsic factors which may determine the rate of increase in cell volume is outlined in terms of the uptake of nutrients. The limiting factors include the available surface for nutrient exchange, the supply of nutrients, the effect of local pH and the influence of stimulatory hormones and growth factors and inhibitors. It may be that one of the features of transformed cells is their constitutively greater nutrient transport with a consequent reduction in sensitivity to extrinsic controlling factors.

Such an alteration would favour growth to a larger population size and could be one of the modes of production of benign tumours. The expression of malignant characteristics seems to involve other changes in cellular behaviour, the nature of which remain obscure.

## Acknowledgements

I thank Mrs. *A. Latter* and Miss *C. Johnson* for help in preparing the manuscript and Mrs. *M. Adams* for its careful transcription. Many of the views expressed in this chapter were developed in discussion with Dr. *P.M. Sutton*. I am grateful to the National Foundation for Cancer Research for financial support.

## References

1. Abell, C.W.; Monahan, T.M.: The role of adenosine 3′, 5′-cyclic monophosphate in the regulation of mammalian cell division. J. Cell Biol. *59:* 549–558 (1973).
2. Abercrombie, M.: Contact inhibition in tissue culture. In Vitro *6:* 128–142 (1970).
3. Abercrombie, M.; Heaysman J.E.M.: Observations on the social behaviour of cells in tissue culture. II. Monolayering of fibroblasts. Expl Cell Res. *6:* 293–306 (1954).
4. Abercrombie, M.; Heaysman, J.E.M.; Karthauser, H.M.: Social behaviour of cells in tissue culture. III. Mutual influence of sarcoma cells and fibroblasts. Expl Cell Res. *13:* 276–291 (1957).
5. Abercrombie, M.; Heaysman, J.E.M.; Pegrum, S.M.: The locomotion of fibroblasts in culture. 3. Movements of particles on the dorsal surface of the leading lamella. Expl Cell Res. *62:* 389–398 (1970).
6. Abercrombie, M.; Lamont, D.M.; Stephenson, E.M.: The monolayering in tissue culture of fibroblasts from different sources. Proc. R. Soc. *170:* 349–360 (1968).
7. Absher, P.M.; Absher, R.G.: Clonal variation and ageing of diploid fibroblasts. Expl Cell Res. *103:* 247–255 (1976).

8   Absher, P.M.; Absher, R.G.; Barnes, W.D.: Genealogies of clones of diploid fibroblasts. Expl Cell Res. *88:* 95–104 (1974).
9   Allen, C.; Saba, T.M. and Molnar, J.: Isolation, purification and characterisation of opsonic protein. J. reticuloendoth. Soc. *13:* 410–423 (1973).
10  Antoniades, H.N.; Stathakos, D.; Scher, C.D.: Isolation of a cationic polypeptide from human serum that stimulates proliferation of 3T3 cells. Proc. natn. Acad. Sci. USA *72:* 2635–2639 (1975).
11  Armelin, H.A.: Pituitary extracts and steroid hormones in the control of 3T3 cell growth. Proc. natn. Acad. Sci. USA *70:* 2702–2706 (1973).
12  Balázs, A.: Control of cell proliferation by endogenous inhibitors (Elsevier/North-Holland, Amsterdam 1979)
13  Barnes, D.; Sato, G.: Growth of a human mammary tumour cell line in a serum-free medium. Nature, Lond. *281:* 388–389 (1979).
14  Bartlett, M.S.: Equations and models in population change; in Bartlett, Hiorns, Mathematical theory of the dynamics of biological populations, pp. 5–21 (Academic Press, London 1973).
15  Baserga, R.: Biochemistry of the cell cycle: a review. Cell Tissue Kinet. *1:* 167–191 (1968).
16  Baserga, R.; Estensen, R.D.; Petersen, R.O.: Inhibition of DNA synthesis in Ehrlich ascites cells by actinomycin D. II. The presynthetic block in the cell cycle. Proc. natn. Acad. Sci. USA *54:* 1141–1148 (1965).
17  Baserga, R.; Ming, P.M.L.; Tsutsui, Y.; Whelly, S.; Chang, H.; Rossini, M.; Huang, C.H.: Nuclear control of cell proliferation; in Brinkley, Porter, International cell biology, pp. 409–413 (Rockefeller University Press, New York 1977).
18  Bateman, A.E.: Cell specificity of chalone-type inhibitors of DNA synthesis released by blood leucocytes and erythrocytes. Cell Tiss. Kinet. *7:* 451–461 (1974).
19  Benbow, R.M.; Joenje, H.; White, S.H.; Breaux, C.B.; Krauss, M.R.; Ford, C.C.; Laskey, R.A.: Cytoplasmic control of nuclear DNA replication in *Xenopus laevis;* in Brinkley, Porter, International cell biology, pp. 453–463 (Rockefeller University Press, New York 1977).
20  Bernard, C.: Sur les phénomènes de la vie (Paris 1878).
21  Von Bertalanffy, L.: Principles and theory of growth; in Nowinski, Fundamental aspects of normal and malignant growth, pp. 137–259 (Elsevier, Amsterdam 1960).
22  Bertazzoni, U.; Steffenini, M.; Noy, G.; Guilotto, E.; Nuzzo, F.; Falaschi, A.; Spadari, S.: Variations of DNA-polymerases $\alpha$ and $\beta$ during prolonged stimulation of human lymphocytes. Proc. natn. Acad. Sci. USA *73:* 785–789 (1976).
23  Bloch, D.P.; MacQuigg, R.A.; Brach, S.D.; Wu, J.R.: The synthesis of deoxyribonucleic acid and histone in onion root meristem. J. Cell Biol. *33:* 451–468 (1967).
24  Bombik, B.M.; Burger, M.M.: cAMP and the cell cycle: inhibition of growth stimulation. Expl Cell Res. *80:* 88–94 (1973).
25  Bourne, H.R.; Coffino, P.; Tomkins, G.M.: Somatic genetic analysis of cyclic AMP action: characterisation of unresponsive mutants. J. cell. Physiol. *85:* 611–626 (1975).
26  Bresciani, F.: DNA synthesis in alveolar cells of the mammary gland: acceleration by ovarian hormones. Science, N.Y. *146:* 653–655 (1964).
27  Bresciani, F.; Puca, G.A.; Nola, E.; Sica, V.: Early stages in estrogen control of cell proliferation; in Clarkson, Baserga, Control of proliferation of animal cells, vol. I, pp. 67–83 (Cold Spring Harbor Laboratory, New York 1974).

28   Brewer, E.N.; Rusch, H.P.: Effect of elevated temperature shocks on mitosis and on the initiation of DNA replication in *Physarum polycephalum.* Expl Cell Res. *49:* 79–86 (1968).
29   Brooks, R.F.: Regulation of the fibroblast cell cycle by serum. Nature, Lond. *260:* 248–250 (1976).
30   Brooks, R.F.: Cell cycle (Cambridge University Press, London, in press, 1980).
31   Brunner, M.: Regulation of DNA synthesis by amino acid limitation. Cancer Res. *33:* 29–32 (1973).
32   Bryant, P.J.: Cell lineage relationships in the imaginal wing disc of *Drosophila melanogaster.* Devl. Biol. *22:* 389–411 (1970).
33   Burch, P.R.J.: New approach to cancer. Nature, Lond. *225:* 512–516 (1970).
34   Burns, F.J.; Tannock, I.F.: On the existence of a $G_0$-phase in the cell cycle. Cell Tiss. Kinet. *3:* 321–334 (1970).
35   Cairns, J.: Cancer: Science and society (Freeman, San Francisco 1978).
36   Calothy, G.; Croce, C.; Defendi, V.; Kaprowski, H.; Eagle, H.: Effect of environmental pH on rescue of Simian virus 40. Proc. natn. Acad. Sci. USA *70:* 366–368 (1973).
37   Canagaratna, M.C.P.; Riley, P.A.: The pattern of density dependent growth inhibition in murine fibroblasts. J. cell. Physiol. *85:* 271–282 (1975).
38   Canagaratna, M.C.P.; Chapman, R.E.; Ehrlich, E.; Sutton, P.M.; Riley, P.A.: Evidence for long-range effects in density-dependent inhibition of proliferation (DDIP) in 3T3 cells. Differentiation *9:* 157–160 (1977).
39   Cannon, W.B.: The wisdom of the body. (Norton, New York 1939).
40   Carter, S.B.: Haptotaxis and the mechanism of cell motility. Nature, Lond. *213:* 256–260 (1967).
41   Caspersson, T.; Farber, S.; Foley, G.E.; Killander, D.; Zetterberg, A.: Cytochemical evaluation of metabolic inhibitors in cell cultures. Expl. Cell. Res. *39:* 365–385 (1965).
42   Castor, L.N.: Control of division by cell contact and serum concentration in cultures of 3T3 cells. Expl. Cell Res. *68:* 17–24 (1971).
43   Castor, L.N.: Contact inhibition of cell division and cell movement. J. invest. Derm. *59:* 27–34 (1972).
44   Ceccarini, C.; Eagle, H.: Induction and reversal of contact inhibition of growth by pH modification. Nature new Biol. *233:* 271–273 (1971).
45   Ceccarini, C.; Eagle, H.: pH as a determinant of cellular growth and contact inhibition. Proc. natn. Acad. Sci. USA *68:* 229–233 (1971).
46   Chapman, R.E.; Marsh, E.; Sutton, P.M.; Riley, P.A.: Studies of long-range density effects on the proliferation of 3T3 and RLCW cells in recirculated medium. Differentiation *10:* 159–164 (1978).
47   Chen, L.B.; Gudor, R.C.; Sun, T.T.; Chen, A.B.; Mosesson, M.W.: Control of a cell surface major glycoprotein by epidermal growth factor. Science, N.Y. *197:* 776–778 (1977).
48   Clark, J.; Jones, K.L.; Gospodarowicz, D.; Sato, G.H.: Growth response to hormones by a new rat ovary cell line. Nature new Biol. *236:* 180–181 (1972).
49   Clarke, G.D.; Smith, C.: The response of normal and polyoma virus-transformed BHK/21 cells to exogenous purines. J. cell. Physiol. *81:* 125–132 (1973).
50   Clutterbuck, A.J.: Cell volume per nucleus in haploid and diploid strains of *Aspergillus nidulans.* J. gen. Microbiol. *55:* 291–299 (1969).

51  Coffino, P.; Gray, J.W.; Tomkins, G.M.: Cyclic AMP, a non-essential regulator of the cell cycle. Proc. natn. Acad. Sci. USA 72: 878–882 (1975).
52  Cohen, S.; Taylor, J.M.: Epidermal growth factor. I. Chemical and biological characterisation Recent Prog. Horm. Res. 30: 533–550 (1974).
53  Collins, J.M.: Deoxyribonucleic acid structure in human diploid fibroblasts stimulated to proliferate. J. biol. Chem. 252: 141–147 (1977).
54  Crick, F.H.C.; Lawrence, P.A.: Compartments and polyclones in insect development: clones made in early development keep within certain fixed boundaries in insect epithelium. Science, N.Y. 189: 340–347 (1975).
55  Crile, G.: A speculative review of the role of endocrine imbalances in the genesis of certain cancers and degenerative diseases. J. natn. Cancer Inst. 20: 229–243 (1958).
56  Cuatrecasas, P.: Properties of the insulin receptor isolated from liver and fat cell membranes. J. biol. Chem. 247: 1980–1991 (1972).
57  Culp, L.A.: Molecular composition and origin of substrate-attached material from normal and virus-transformed cells. J. supromol. Struct. 5: 239–255 (1976).
58  Cummins, J.E.; Brewer, E.N.; Rusch, H.P.: The effect of actidione on mitosis in the slime mold *Physarum polycephalum*. J. Cell Biol. 27: 337–341 (1965).
59  Cummins, J.E.; Blomquist, J.C.; Rusch, H.P.: Anaphase delay after inhibition of protein synthesis between late prophase and prometaphase. Science, N.Y. 154: 1343–1344 (1966).
60  Cummins, J.E.; Rusch, H.P.: Limited DNA synthesis in the absence of protein synthesis in *Physarum polycephalum*. J. Cell Biol. 31: 577–583 (1966).
61  Curtis, A.S.G.: The cell surface (Academic Press, London 1967).
62  Damluji, R.; Riley, P.A.: On the role of calcium in adhesion of cells to solid substrates. Expl Cell Biol. 47: 226–237 (1979).
63  Damluji, R.; Riley, P.A.: Time dependency of a critical effect of EGTA on DNA synthesis in cultures of Swiss 3T3 cells. Expl Cell Biol. 47: 446–453 (1979).
64  De Terra, N.: Macronuclear DNA synthesis in *Stentor*; regulation by a cytoplasmic initiator. Proc. natn. Acad. Sci. USA 57: 607–614 (1967).
65  Devi, V.R.; Guttes, E.; Guttes, S.: Effects of ultraviolet light on mitosis in *Physarum polycephalum*. Expl Cell. Res. 50: 589–598 (1968).
66  Di Pasquale, A.: Locomotory activity of epithelial cells in culture. Expl Cell Res. 94: 191–215 (1975).
66a Di Pasquale, A.; Bell, P.B., Jr.: The upper cell surface: its inability to support active cell movement in culture. J. Cell Biol. 62: 198–214 (1974).
67  Donachie, W.D.: Relationship between cell size and the time of initiation of DNA replication. Nature, Lond. 219: 1077–1079 (1968).
68  Dulbecco, R.; Elkington, J.: Induction of growth in resting fibroblastic cell cultures by calcium. Proc. natn. Acad. Sci. USA 72: 1584–1588 (1975).
69  Eagle, H.: Buffer combinations for mammalian cell culture. Science, N.Y. 174: 500–503 (1971).
70  Eagle, H.: The effect of environmental pH on growth of normal and malignant cells. J. cell. Physiol. 82: 1–8 (1973).
71  Evian, D.; Anderson, W.B.: Gonadotropin stimulation of cyclic AMP levels in Chinese hamster ovary cells in culture. J. cell. Physiol. 99: 153–158 (1979).
72  Everhart, L.P.; Prescott, D.M.: Reversible arrest of Chinese hamster cells in GI by partial deprivation of leucine. Expl Cell Res. 75: 170–174 (1972).

73   Fantes, P.A.; Grant, W.D.; Pritchard, R.H.; Sudbery, P.E.; Wheals, A.E.: The regulation of cell size and the control of mitosis. J. theor. Biol. *50:* 213–244 (1975).
74   Fantes, P.; Nurse, P.: Cell cycle (Cambridge University Press, London, in press, 1980).
75   Feller, W.: Die Grundlagen der Volterraschen Theorie des Kampfes ums Dasein in Wahrscheinlichkeitstheoretischer Behandlung. Acta biotheor. *5:* 11–40 (1939).
76   Folkman, J.; Greenspan, H.P.: Influence of geometry on control of cell growth. Biochim. biophys. Acta *417:* 211–236 (1975).
77   Folkman, J.; Moscona, A.: Role of cell shape in growth control. Nature, Lond. *273:* 345–349 (1978).
78   Folks, J.L.; Chhikara, R.S.: The inverse Gaussian distribution and its statistical application – a review. J. r. statist. Soc. *B 40:* 263–289 (1978).
79   Fournier, R.E.; Pardee, A.B.: Cell cycle studies of mononucleate and cythchalasin-B-induced binucleate fibroblasts. Proc. natn. Acad. Sci. USA *72:* 869–873 (1975).
80   Frazier, E.A.J.: DNA synthesis following gross alterations of the nucleocytoplasmic ratio in the ciliate *Stentor coeruleus.* Devl. Biol. *34:* 77–92 (1973).
81   Froehlich, J.E.; Anastassiades, T.P.: Role of pH in fibroblast proliferation. J. cell. Physiol. *84:* 253–260 (1974).
82   Gahmberg, C.G.; Hakamori, S.I.: Altered growth behaviour of malignant cells associated with changes in externally labelled glycoprotein and glycolipid. Proc. natn. Acad. Sci. USA *70:* 3329–3333 (1973).
83   Gail, M.H.; Boone, C.W.: Density inhibition of motility in 3T3 fibroblasts and their SV40 transformants. Expl Cell Res. *64:* 156–162 (1971).
84   Gail, M.H.; Scher, C.D.; Boone, C.W.: Dissociation of cell motility from cell proliferation in BALB/c-3T3 fibroblasts. Expl Cell Res. *70:* 439–443 (1972).
85   Galvasig, G.; Bootsma, D.: Synchronisation of mammalian cells in vitro by inhibition of the DNA synthesis. II. Population dynamics. Expl Cell Res. *41:* 438–451 (1966).
86   Garcia-Bellido, A.: Cell affinities in antennal homoeotic mutants of *Drosophila melanogaster.* Genetics *59:* 487–499 (1968).
87   Gause, G.F.: The struggle for existence (Williams & Wilkins, Baltimore 1934).
88   Gerson, D.F.: Intracellular pH and the mitotic cycle in *Physarum* and mammalian cells; in Jeks, Cameron, Padilla, Zimmerman, Cell cycle regulation, pp. 105–131 (Academic Press, London 1978).
89   Gerson, D.F.; Burton, A.C.: The relation of cycling of intracellular pH to mitosis in the acellular slime mould *Physarum polycephalum.* J. cell. Physiol. *91:* 297–303 (1977).
90   Goldberg, N.D.; Haddox, M.K.; Dunham, E.; Lopez, C.; Madden, J.W.: The Yin-Yang hypothesis of biological control: opposing influences of cyclic GMP and cyclic AMP in the regulation of cell proliferation and other processes; in Clarkson, Baserga, Control of proliferation in animal cells, vol. I, pp. 609–625 (Cold Spring Harbour Laboratory, New York 1974).
91   Gompertz, B.: On the nature of the function expressive of the law of human mortality. Phil. Trans. R. Soc. *1825:* 513–585 (1825).
92   Gordon, A.S.; Zanjani, E.D.; Gidaro, A.S.; Kuna, R.A.: Erythropoieton: the humoral regulator of erythropoiesis; in LoBue, Gordon, Humoral control of growth and differentiation, vol. I, pp. 25–49 (Academic Press, New York 1973).
93   Gospodarowicz, D.; Gospodarowicz, F.: The morphological transformation and inhibition of growth of bovine luteal cells in tissue culture induced by leuteinising hormone and cyclic AMP. Endocrinology *96:* 458–467 (1975).

94 Gospodarowicz, D.; Moran, J.: Optimal conditions for the study of growth control of BALB/c 3T3 fibroblasts. Expl Cell Res. *90:* 279–284 (1975).
95 Gospodarowicz, D.; Greene, G.; Moran, J.: Fibroblast growth factor can substitute for platelet factor to sustain growth of BALB/c 3T3 cells in the presence of plasma. Biochem. biophys. Res. Commun. *65:* 779–787 (1975).
96 Graham, C.F.; Arms, K.; Gurdon, J.B.: The induction of DNA synthesis by frog egg cytoplasm. Devl Biol. *14:* 349–381 (1966).
97 Graves, J.A.: DNA synthesis in heterokaryons formed by fusion of mammalian cells from different species. Expl Cell Res. *72:* 393–403 (1972).
98 Griffiths, J.B.: The effect of medium changes on the growth and metabolism of the human diploid cells WI-38. J. Cell Sci. *8:* 43–52 (1971).
99 Griffiths, J.B.: The effect of cell population density on nutrient uptake and cell metabolism. J. Cell Sci. *10:* 515–524 (1972).
100 Griffiths, J.B.: Role of serum, insulin and amino acid concentration in contact inhibition of growth of human cells in culture. Expl Cell Res. *75:* 47–56 (1972).
101 Grinnell, F.: The serum dependence of baby hamster kidney cell attachment to a substratum. Expl Cell Res. *97:* 265–274 (1976).
102 Grinnell, F.; Hays, D.G.; Minter, D.: Cell adhesion and spreading factor. Partial purification and properties. Expl Cell Res. *110:* 175-190 (1977).
103 Haldane, J.B.S.: A mathematical theory of natural and artificial selection. Part I. Trans. Camb. phil. Soc. *23:* 19–45 (1924).
104 Harel, L.; Jullien, M.: Evaluation of proximity inhibition of DNA synthesis in 3T3 cells. J. cell. Physiol. *88:* 253–254 (1976).
105 Harris, A.K.: Contact inhibition of cell locomotion; in Cox, cell communication, pp. 147–185 (Wiley, New York 1974).
106 Hartmann, M.: Über experimentelle Unsterblichkeit von Protozoen-Individuen. Ersatz der Fortpflanzung von *Amoeba proteus* durch fortgesetzte Regeneration. Zool. Jb. Abt. allg. Zool. *45:* 973–987 (1928).
107 Hayflick, L.: The limited in vitro lifetime of human diploid cell strains. Expl Cell Res. *37:* 614–636 (1965).
108 Heath, J.P.; Dunn, G.A.: Cell to substratum contacts of chick fibroblasts and their relation to the microfilament system. A correlated interference-reflexion and high voltage electron-microscope study. J. Cell Sci. *29:* 197–212 (1978).
109 Heaysman, J.E.M.: Contact inhibition of locomotion: a re-appraisal. Int. Rev. Cytol. *55:* 49–66 (1978).
110 Heaysman, J.E.M.; Turin, L.: Interactions between living and zinc-fixed cells in culture. Expl Cell Res. *101:* 419–422 (1976).
111 Hintz, R.L.; Clemmons, D.R.; Underwood, L.E.; Van Wyk, J.J.: Competitive binding of somatomedin to the insulin receptors of adipocytes, chondrocytes and liver membranes. Proc. natn. Acad. Sci. USA *69:* 2351–2353 (1972).
112 Hollenberg, M.D.; Cuatrecasas, P.: Insulin and epidermal growth factor. J. biol. Chem. *250:* 3845–3853 (1975).
113 Holley, R.W.: Serum factors and growth control; in Clarkson, Baserga, Control of proliferation in animal cells, vol. I, pp. 13–18 (Cold Spring Harbor Laboratory, New York 1974).
114 Holley, R.W.: Control of growth of mammalian cells in culture. Nature, Lond. *258:* 487–490 (1975).

115 Holley, R.W.; Kiernan, J.A.: Contact inhibition of cell division in 3T3 cells. Proc. natn. Acad. Sci. USA 60: 300–304 (1968).
116 Honegger, P.; Lenoir, D.; Faroud, P.: Growth and differentiation of aggregating fetal brain cells in a serum-free defined medium. Nature, Lond. 282: 305–307 (1979).
117 Hovi, T.; Vaheri, A.: Cyclic AMP and cyclic GMP enhance growth of chick embryo fibroblasts. Nature new Biol. 245: 175–177 (1973).
118 Hovi, T.; Keski-Oja, J.; Vaheri, A.: Growth control in chick embryo fibroblasts: no evidence for a specific role for cyclic purine nucleotides. Cell 2: 235–240 (1974).
119 Howard, A.; Pelc, S.R.: Nuclear incorporation of $P^{32}$ as demonstrated by autoradiographs. Heredity 5: 158 (1951).
120 Hutchinson, G.E.: Concluding remarks. Cold Spring Harb. Symp. quant. Biol. 22: 415–427 (1957).
121 Hynes, R.O.; Bye, J.M.: Density and cell cycle dependence of cell surface proteins in hamster fibroblasts. Cell 3: 113–120 (1974).
122 Jimenez de Asua, L.; Clingan, D.; Rudland, P.S.: Initiation of cell proliferation in cultured mouse fibroblasts by prostaglandin $F_2\alpha$. Proc. natn. Acad. Sci. USA 72: 2724–2728 (1975).
123 Johnson, J.D.; Epel, D.; Paul, M.: Intracellular pH and activation of sea urchin eggs after fertilisation. Nature, Lond. 262: 661–664 (1976).
124 Kaczanowski, A.: Mitosis and polyploidy in nuclei of *Opalina ranarum*. Experientia 24: 846–847 (1968).
125 Kemp, H.H.; Gey, G.D.: Changes in serum proteins during growth of malignant cells in vitro. Proc. Soc. exp. Biol. Med. 94: 205–208 (1957).
126 Killander, D.; Zetterberg, A.: Quantitative cytochemical studies on interphase growth. I. Determination of DNA, RNA and mass content of age determined mouse fibroblasts in vitro and of intercellular variation in generation time. Expl Cell Res. 38: 272–284 (1965).
127 Killander, D.; Zetterberg, A.: A quantitative cytochemical investigation of the relationship between cell mass and initiation of DNA synthesis in mouse fibroblasts in vitro. Expl Cell Res. 40: 12–20 (1965).
128 Kim, J.H.; Gelbard, A.S.; Perez, A.G.: Inhibition of DNA synthesis by actinomycin D and cycloheximide in synchronised HeLa cells. Expl Cell Res. 53: 478–487 (1968).
129 Klebe, R.J.: Isolation of a collagen-dependent cell attachment factor. Nature, Lond. 250: 248–251 (1974).
130 Koch, A.L.: Does the variability of the cell cycle result from one or many chance events? Nature, Lond. 286: 80–82 (1980).
131 Koch, K.S.; Lieffert, H.L.; Moran, T.: Hepatic proliferation control by purines, hormones and nutrients; in Fishman, Onco-developmental gene expression, pp. 21–33 (Academic Press, New York 1976).
132 Koch, A.L.; Schaechter, M.: A model for statistics of the cell division process. J. gen. Microbiol. 29: 435–454 (1962).
133 Kolodny, G.M.; Gross, P.R.: A simple mechanical method for the efficient release of contact inhibition. Expl Cell Res. 57: 423–432 (1969).
134 Kothari, I.R.; Martin, G.C.; Reilly, P.J.; Martin, P.J.; Eakman, J.M.: Estimation of parameters in population models for *Schizo-saccharomyces pombe* from chemostat data Biotech. Bioengng 14: 915–938 (1972).
135 Kuusela, P.; Ruoslamti, E.; Engvall, E.; Vaheri, A.: Immunological interspecies cross-

reactions of fibroblast surface antigen (fibronectin). Immunochemistry *13:* 639–642 (1976).

136 Laurence, E.B.: The significance of chalones in epidermal growth; in Spearman, Riley, The skin of vertebrates, pp. 139–149 (Academic Press, London 1980).

137 Leffert, H.L.: Growth control of differentiated fetal rat hepatocytes in primary monolayer culture. VII. Hormonal control of DNA synthesis and its possible significance to the problem of liver regeneration. J. Cell Biol. *62:* 792–801 (1974).

138 Levine, S.; Pictet, R.; Rutter, W.J.: Control of cell proliferation and cytodifferentiation by a factor reacting with the cell surface. Nature new Biol. *246:* 49–52 (1973).

139 Ley, K.D.; Tobey, R.A.: Regulation of initiation of DNA synthesis in Chinese hamster cells. I. Production of stable, reversible, G1-arrested populations in suspension culture. J. Cell Biol. *46:* 151–157 (1970).

140 Ley, M.D.: Detection of G1 proteins in Chinese hamster cells synchronised by isoleucine deprivation or mitotic selection. J. Cell Biol. *66:* 95–101 (1975).

141 Liao, S.: Cellular receptors and mechanisms of action of steroid hormones. Int. Rev. Cytol. *41:* 87–172 (1975).

142 Lie, S.O.; McKusick, V.A.; Neufeld, E.F.: Stimulation of genetic mucopolysaccharidoses in normal human fibroblasts by alteration of pH of the medium. Proc. natn. Acad. Sci. USA *69:* 2361–2362 (1972).

143 Linnaeus, C.: Oratio de telluris Habitabilis Incremento. Fundam. Botanic § 132 (1743); 59–60 in Amoenitates academicae, 3rd ed., vol. 2, pp. 446–447 (Palm, Erlangen 1787).

144 Littlefield, J.W.: DNA synthesis in partially synchronised L cells. Expl Cell Res. *26:* 318–326 (1962).

145 Lotka, A.J.: Elements of mathematical biology (Williams & Wilkins, Baltimore 1925).

146 Lotka, A.J.: The growth of mixed populations: two species competing for a common food supply. J. Wash. Acad. Sci. *22:* 461–469 (1932).

147 Mac Vittie, T.J.; McCarthy, K.F.: The influence of granulocytic inhibitor(s) on hematopoiesis in an in vivo culture system. Cell Tiss. Kinet. *8:* 553–559 (1975).

148 McLimans, W.F.; Blumenson, L.E.; Tunnam, K.V.: Kinetics of gas diffusion in mammalian cell culture systems. II. Theor. Biotech. Bioengng. *10:* 741–763 (1968).

149 Malthus, T.R.: An essay on the principle of population as it affects the future improvement of society, with remarks on the speculations of Mr. Godwin, M. Condorçet, and other writers (published anonymously, 1798).

150 Manfredi, F.: Calculation of total body intracellular pH in normal human subjects from the distribution of 5, 5-di-methyl-2, 4-oxazolidinedione (DMO). J. Lab. clin. Med. *61:* 1005–1014 (1963).

151 Marciani, D.J.; Bader, J.P.: Polypeptide composition of cell membranes from chick embryo fibroblasts transformed by Rous sarcoma virus. Biochim. biophys. Acta *401:* 386–398 (1975).

152 Martz, E.; Phillips, H.M.; Steinberg, M.S.: Contact inhibitions of overlapping and differential cell adhesion: a sufficient model for the control of certain cell culture morphologies. J. Cell Sci. *16:* 401–409 (1974).

153 Matsuzawa, H.; Nirenberg, M.: Receptor-mediated shifts in cGMP and cAMP levels in neuroblastoma cells. Proc. natn. Acad. Sci. USA *72:* 3472–3476 (1975).

154 Mauck, J.C.; Green, H.: Regulation of RNA synthesis in fibroblasts during transition from resting to growing state. Proc. natn. Acad. Sci. USA *70:* 2819–2822 (1973).

155 May, R.W.: Stability and complexity in model eco-systems (Princeton University Press, Princeton 1973).
156 Maynard-Smith, J.: Mathematical ideas in biology, pp. 46–50 (Cambridge University Press, London 1968).
157 Mayr, O.: The origins of feedback control (MIT Press, Cambridge 1970).
158 Menyhárt, J.; Marcsek, Z.; Menczel, L.; Gróf, J.: Stimulation and inhibition of DNA synthesis by polypeptides isolated from intact and regenerating rat livers. FEBS Abstr. *1975:* 1611.
159 Messmer, T.O.: A comparison of the interaction of Chinese hamster fibroblasts with human and bovine transferrin, and conalbumin (chick-egg transferrin). Biochim. biophys. Acta *320:* 663–670 (1973).
160 Metcalf, D.: Regulation by colony-stimulating factor of granulocyte and macrophage colony formation in vitro by normal and leukaemic cells; in Clarkson, Baserga, Control of proliferation in animal cells, vol. I, pp. 887–905 (Cold Spring Harbour Laboratory, New York, 1974).
161 Middleton, C.A.: Contact inhibition of locomotion in cultures of pigmented retina epithelium. Expl Cell Res. *70:* 91–96 (1972).
162 Middleton, C.A.: Contact-induced spreading is a new phenomenon depending on cell-cell contact. Nature, Lond. *259:* 311–313 (1976).
163 Minor, P.D.; Smith, J.A.: Explanation of degree of correlation of sibling generation times in animal cells. Nature, Lond. *248:* 241–243 (1974).
164 Mitchison, J.M.: The biology of the cell cycle (Cambridge University Press, London 1971).
165 Mitchison, J.M.; Creanor, J.: Introduction synchrony in the fission yeast. *Schizosaccharomyces pombe.* Expl Cell Res. *67:* 368–374 (1971).
166 Moens, W.; Wokaer, A.; Kram, R.: Cyclic AMP and cyclic GMP concentrations in serum and density-restricted fibroblast cultures. Proc. natn. Acad. Sci. USA *72:* 1063–1067 (1975).
167 Morris, V.B.; Cowan, R.; Culpin, D.: Variability of cell cycle times measured *in vivo* in embryonic chick retina by continuous labelling with BUdR. Nature, Lond. *280:* 68–70 (1979).
168 Morrison, P.; Edsall, R.; Miller, S.G.: Preparation and properties of serum and plasma proteins. XVIII. The separation of purified fibrinogen from fraction I of human plasma. J. Am. chem. Soc. *70:* 3103–3108 (1948).
169 Mosher, D.F.; Vaheri, A.: Thrombin stimulates the production and release of a major surface-associated glycoprotein (fibronectin) in cultures of human fibroblasts. Expl Cell Res. *112:* 323–334 (1978).
170 Mueller, G.C.; Kajiwara, K.; Ichikawa, A.; Planck, S.: Molecular events in the replication of DNA and chromation; in Brinckley, Porter, International cell biology pp. 430–435 (Rockefeller University Press, New York 1977).
171 Nägeli, C.: Verdrängung der Pflanzenformen durch ihre Mitbewerber. Sber. Math. Phys. Klasse Akad. Wiss., München *4:* 109–164 (1874).
172 Oldfield, F.E.: Orientation behavior of chick leucocytes in tissue culture and their interactions with fibroblasts. Expl Cell Res. *30:* 125–138 (1963).
173 O'Neill, C.M.; Riddle, P.N.; Jordan, P.W.: The relation between surface area and anchorage dependence of growth in hamster and mouse fibroblasts. Cell *16:* 909–918 (1979).

174 Osgood, E.E.: A unifying concept of the etiology of leukaemias, lymphomas and cancers. J. natn. Cancer Inst. *18:* 155–166 (1957).
175 Pavkovits, W.R.: Control of gramlocyte production: separation and chemical identification of a specific inhibitor (chalone). Cell Tiss. Kinet *4:* 539–547 (1971).
176 Pawalek, J.; Halaban, R.; Christie, G.: Melanoma cells which require cyclic AMP for growth. Nature, Lond. *258:* 539–540 (1975).
177 Pearl, R.; Read, L.J.: On the rate of growth of the population of the United States since 1790 and its mathematical representation. Proc. natn. Acad. Sci. USA *6:* 275–288 (1920).
178 Pearlstein, E.; Waterfield, M.D.: Metabolic studies on $^{125}$I-labelled baby hamster kidney cell plasma membranes. Biochim. biophys. Acta *362:* 1–12 (1974).
179 Pena, S.D.J.; Hughes, R.C.: Fibronectin – plasma membrane interactions in the adhesion and spreading of hamster fibroblasts. Nature, Lond. *276:* 80–83 (1978).
180 Pickart, L.; Thaler, M.M.: Tripeptide in human serum which prolongs survival of normal liver cells and stimulates growth in neoplastic liver. Nature, Lond. *243:* 85–87 (1973).
181 Pollard, J.W.; Stanners, C.P.: Characterisation of cell livers showing growth control isolated from both the mild type and a leucyl-tRNA synthetase mutant of Chinese hamster ovary cells. J. cell. Physiol. *98:* 571–586 (1979).
182 Prandtl, L.: Über Flüssigkeitsbewegungen bei sehr kleiner Reibung. Proc. 3rd Int. Mathematics Congress, Heidelberg 1904.
183 Prescott, D.M.: Relation between cell growth and cell division. II. The effect of cell size on cell growth rate and generation time in *Amoeba proteus*. Expl Cell Res. *11:* 86–94 (1956).
184 Prescott, D.M.: Relation between cell growth and cell division. III. Changes in nuclear column and growth rate and prevention of cell division in *Amoeba proteus* resulting from cytoplasmic amputations. Expl Cell Res. *11:* 94–98 (1956).
185 Prescott, D.M.: The cell cycle and the control of cellular reproduction. Adv. Genet. *18:* 99–177 (1976).
186 Pritchard, R.H.; Lark, K.G.: Introduction of replication by thymine starvation at the chromosome origin in *Escherichia coli*. J. molec. Biol. *9:* 288–307 (1964).
187 Ramachandran, J.; Suyama, A.T.: Inhibition of replication of normal adrenocortical cells in culture by adrenocorticotropin. Proc. natn. Acad. Sci. USA *72:* 113–117 (1975).
188 Rao, P.N.; Johnson, R.T.: Mammalian cell fusion: studies on the regulation of DNA synthesis and mitosis. Nature, Lond. *225:* 159–164 (1970).
189 Rao, P.N.; Smith, M.L.: Regulation of DNA synthesis in cytochalasin B(CB)-induced binucleate HeLa cells. Expl Cell Res. *103:* 213–218 (1976).
190 Rao, P.N.; Sunkara, P.S.: Cell fusion and regulation of DNA synthesis, pp. 133–147, in Jeter, Cameron, Padilla, Zimmerman, Cell cycle regulation, (Academic Press, London 1978).
191 Rees, D.A.; Lloyd, C.W.; Thom, D.: Control of grip and stick in cell adhesion through lateral relationships of membrane glycoproteins. Nature, Lond. *267:* 124–128 (1977).
192 Ribbert, H.: Über die Histogenese und das Wachstum des Carcinoms. Virchows Arch. Abt. A Path. Anat. *141:* 153–177 (1895).
193 Riley, P.A.: The influence of differentiation on mitotic regulation: some theoretical aspects; in Harris, Allin, Viza, Cell differentiation, pp. 288–294 (Munksgaard, Copenhagen 1972).

194 Riley, P.A.: The effect on cell proliferation of reduced substrate adhesiveness. Cell Differ. *3:* 233–238 (1974).
195 Riley, P.A.: Growth inhibition in normal mammalian melanocytes in vitro. Br. J. Derm. *92:* 291–304 (1975).
196 Riley, P.A.: A theory of cellular senescence based ond Darwinian principles in the light of Linnaeus; in Spearman, Riley, The skin of vertebrates, pp. 111–125 (Academic Press, London 1981).
197 Riley, P.A.; Canagaratna, M.C.P.: Cell interaction in cultured hepatocytes. Panminerva med. *18:* 492–498 (1976).
198 Riley, P.A.; Dean, R.T.: Phagocytosis of latex particles in relation to the cell cycle in 3T3 cells. Expl Cell Biol. *46:* 367–373 (1978).
199 Riley, P.A.; Hola, M.: Clonal differences in generation times of GPK epithelial cells in monolayer culture. Expl Cell Biol. *48:* 310–320 (1980).
200 Riley, P.A.; Hola, M.: Clonal variation in proliferation rate of cultures of GPK cells. Cell Tiss. Kinet. (to be published, 1981).
201 Riley, P.A.; Hola, M.: Variation in intermitotic interval between sister cells related to orientation at mitosis. Eur. J. Cell Biol. *22:* 315 (1980).
202 Riley, P.A.; Latter, T.; Sutton, P.M.: Hormone assays on breast tumour cultures. Lancet *ii:* 818–820 (1973).
203 Riley, P.A.; Latter, A.; Sutton, P.M.: Reversible morphological changes in human breast fibroblasts exposed to stilboestrol. Eur. J. Cancer *14:* 579–586 (1978).
204 Robbins, J.C.; Nicolson, G.L.: Biology of tumours: surfaces, immunology and comparative pathology; in Becker, Cancer: a comprehensive treatise, vol. 4, pp. 3–54 (Plenum Press, New York 1975).
205 Robbins, E.; Scharff, M.D.: The absence of a detectable G1 phase in a cultured strain of Chinese hamster lung cell. J. Cell Biol. *34:* 684–685 (1967).
206 Robbins, P.W.; Wickus, G.G.; Branston, P.E.: The chick fibroblast cell surface after transformation by Rous sarcoma virus. Cold Spring Harb. Symp. quant. Biol. *39:* 1173–1180 (1974).
207 Rodbell, M.: The role of hormone receptors and GTP-regulatory proteins in membrane transduction. Nature, Lond. *284:* 17–22 (1980).
208 Rodbell, M.; Lin, M.C.; Salomon, Y.; Londos, C.; Hawood, J.P.; Martin, B.R.; Rendell, M.; Berman, M.: Role of adenine and guanine nucleotides in the activity and response of adenylate cyclase systems to hormones: evidence for multisite transition states. Adv. cyclic Nucleotide Res. *5:* 3–29 (1975).
209 Ross, R.: The prevention of malaria (Murray, London 1911).
210 Ross, R.; Bornstein, P.: The elastic fiber. I. The separation and partial characterization of its macromolecular components. J. Cell Biol. *40:* 366–381 (1969).
211 Rubin, H.: pH and population density in the regulation of animal cell multiplication. J. Cell Biol. *51:* 686–702 (1971).
212 Rubin, H.: pH, serum and zinc in the regulation of DNA synthesis in cultures of chick embryo cells. J. cell. Physiol. *82:* 231–238 (1973).
213 Rubin, H.; Rein, A.: Proximity effects in the growth of animal cells; in Defendi, Stoker, Growth regulating substances for animal cells in culture, pp. 51–66 (Wistar Institute Press, Philadelphia 1967).
214 Rudland, P.S.; Gospodarowicz, D.; Seifert, W.: Activation of guanyl cyclase and intracellular cyclic GMP by fibroblast growth factor. Nature, Lond. *250:* 741–743 (1974).

215 Ruoslahti, E.; Vaheri, A.: Novel human serum protein from fibroblast plasma membrane. Nature, Lond. *248:* 789–791 (1974).
216 Sachsenmaier, W.; Dönges, K.; Rupff, H.; Czihak, G.: Advanced initiation of synchronous mitoses in *Physarum polycephalum* following UV-irradiation. Z. Naturf. *25:* 866–871 (1970).
217 Schauenstein, E.; Esterbauer, H.; Zollner, H.: Aldehydes in biological systems. Their natural occurrence and biological activities (Pion Press, London 1977).
218 Schutz, L.; Mora, P.A.: The need for direct cell contact in 'contact' inhibition of cell division in culture. J. cell. Physiol. *71:* 1–6 (1968).
219 Shields, R.; Brooks, R.F.; Riddle, P.N.; Capellaro, D.F.; Delia, D.: Cell size, cell cycle and transition probability in mouse fibroblasts. Cell *15:* 469–474 (1978).
220 Short, J.; Brown, R.F.; Husakova, A.; Gilbertson, J.R.; Zemel, R.; Lieberman, I.: Induction of DNA synthesis in the liver of the intact animal. J. biol. Chem. *247:* 1757–1766 (1972).
221 Simard, A.; Corneille, L.; Deschamps, Y.; Verly, W.G.: Inhibition of cell proliferation in the livers of hepatectomised rats by a rabbit hepatic chalone Proc. natn. Acad. Sci. USA *71:* 1763–1766 (1974).
222 Sisken, J.E.; Morasca, L.: Intrapopulation kinetics of the mitotic cycle. J. Cell Biol. *25:* 179–190 (1965).
223 Siskin, J.E.; Kinosita, R.: Timing of DNA synthesis in the mitotic cycle in vitro. J. biophys. biochem. Cytol. *9:* 509–518 (1961).
224 Skellam, J.G.: The formulation and interpretation of mathematical models of diffusionary processes in population biology; in Bartlett, Hiorns, Mathematical theory of the dynamics of biological populations, pp. 63–85 (Academic Press, London 1973).
225 Smith, G.L.; Temin, H.M.: Purified multiplication-stimulating activity from rat liver cell conditioned medium; comparison of biological activities with calf serum, insulin and somatomedin. J. cell. Physiol. *84:* 181–192 (1974).
226 Smith, J.A.; Martin, L.: Do cells cycle? Proc. natn. Acad. Sci. USA *70:* 1263–1267 (1973).
227 Smith, J.R.; Whitney, R.G.: Intraclonal variation in proliferative potential of human diploid fibroblasts: stochastic mechanism for cellular aging. Science, N.Y. *207:* 82–84 (1980).
228 Steinberg, M.S.: Does differential adhesion govern self-assembly processes in histogenesis? Equilibrium configurations and the emergence of a hierarchy among populations of embryonic cells. J. exp. Zool. *173:* 395–433 (1970).
229 Stevens, K.M.: Oxygen requirements for liver cells in vitro. Nature, Lond. *206:* 199 (1965).
230 Stoker, M.P.G.: Role of the diffusion boundary layer in contact inhibition of growth. Nature, Lond. *245:* 200–202 (1973).
231 Stoker, M.P.G.; Piggott, D.: Shaking 3T3 cells: further studies on the diffusion boundary effect. Cell *3:* 207–215 (1974).
232 Stomatoglou, S.C.: Ultrastructural relationship between cell and substrate coats in serum-free and serum-supplemented cultures. J. Ultrastruct. Res. *60:* 203–211 (1977).
233 Swann, M.M.: The control of cell division: a review. II. Special mechanisms. Cancer Res. *18:* 1118–1160 (1958).
234 Szent-Györgyi, A.: Electronic biology and cancer. A new theory of cancer (Dekker, New York 1976).

235 Taylor, A.C.: Responses of cells to pH changes in the medium. J. Cell Biol. *15:* 201–209 (1962).
236 Taylor, E.W.: Control of DNA synthesis in mammalian cells in cultures. Expl Cell Res. *40:* 316–332 (1965).
237 Terasima, T.; Yasukawa, M.: Synthesis of G1 protein preceding DNA synthesis in cultured mammalian cells. Expl. Cell Res. *44:* 669–671 (1966).
238 Thrash, C.R., Cunningham, D.D.: Stimulation of division of density inhibited fibroblasts by glucocorticoids. Nature, Lond. *242:* 399–401 (1973).
239 Thrash, C.R.; Cunningham, D.D.: Growth limitation of 3T3 mouse fibroblasts by available growth surface area and medium components. J. cell. Physiol. *86:* 301–310 (1975).
240 Tisdale, M.J.; Phillips, B.J.: The relationship between the growth characteristics of somatic cell hybrids and their levels of cAMP and activities of adenyl cyclase and cAMP phosphodiesterase. Expl Cell Res. *99:* 63–71 (1976).
241 Tobey, R.A.; Ley, K.D.: Regulation of initiation of DNA synthesis in Chinese hamster cells. I. Production of a stable, reversible G1-arrested population in suspension culture. J. Cell Biol. *46:* 151–157 (1970).
242 Todaro, G.J.; Lazar, G.K.; Green, H.: The initiation of cell division in a contact-inhibited mammalian cell line. J. cell. comp. Physiol. *66:* 325–334 (1965).
243 Tomkins, G.M.: Cycle dependent enzyme induction; in Richmond et al., Mammalian cells: probes and problems, pp. 145–151 (US Energy Research and Development Administration, Washington 1975).
244 Vandermeer, J.H.: Interspecific competition: a new approach to the classical theory. Science, N.Y. *188:* 253–255 (1975).
245 Van Wijk, R.; Van de Poll, K.W.: Variability of cell generation times in a hepatoma cell pedigree. Cell Tiss. Kinet. *12:* 659–663 (1979).
246 Van Wijk, R.; Van de Poll, K.W.; Amesz, W.J.C.; Geilenkirchen, W.L.M.: Studies on the variations in generation times of rat hepatoma cells in culture. Expl Cell Res. *109:* 371–379 (1977).
247 Van Wyk, J.J.; Underwood, L.E.; Hintz, A.L.; Clemmons, D.R.; Voina, S.; Weaver, R.P.: The somatomedins: a family of insulin-like hormones under growth hormone control. Recent Prog. Horm. Res. *30:* 259–318 (1974).
248 Verhulst, P.F.: Notice sur la loi que la population suit dans son accroissement. Comm. math. phys. *10:* 113–115 (1839).
249 Verly, W.G.: The hepatic chalone. Natn. Cancer Inst. Monogr. *38:* 175–184 (1973).
250 Vesely, P.; Weiss, R.A.: Cell locomotion and contact inhibition of normal and neoplastic rat cells. Int. J. Cancer *11:* 64–76 (1973).
251 Volterra, V.: Leçons sur la theorie mathematique de la lutte pour la vie. (Gauthier-Villas, Paris 1931).
252 Weber, T.H.; Goldberg, M.L.: Effect of leukoagglutinating phytohaemagglutinin on cAMP and cGMP levels in lymphocytes. Expl Cell Res. *97:* 432–435 (1976).
253 Weiss, L.: Symposium on adhesion and cohesion (Elsevier, Amsterdam 1962).
254 Weiss, P.; Kavanau, J.L.: A model of growth and growth control in mathematical terms. J. gen. Physiol. *41:* 1–47 (1957).
255 Werrlein, R.J.; Glinos, A.D.: Oxygen microenvironment and respiratory oscillations in cultured mammalian cells Nature, Lond. *251:* 317–319 (1974).
256 Whitfield, J.F.; MacManus, J.P.; Gillan, D.J.: The ability of calcium to change cyclic AMP from a stimulator to an inhibitor of thymic lymphoblast proliferation. J. cell.

Physiol. 81: 241–250 (1973).
257 Whitfield, J.F.; MacManus, J.P.; Boynton, A.L.; Gillan, D.J.; Isaacs, R.J.: Concanavalin A and the inhibition of thymic lymphoblast DNA synthesis and proliferation by a calcium-dependent increase in cyclic GMP level. J. cell. Physiol. 84: 445–458 (1974).
258 Wiener, N.: Cybernetics: or control and communication in the animal and the machine; 2nd ed. (Wiley, New York 1948).
259 Winsor, C.P.: The Gompertz curve as a growth curve. Proc. natn. Acad. Sci. USA 18: 1–8 (1932).
260 Wolff, I.; Timpl, R.; Pecker, I.; Steffen, C.: A two-component system of human serum agglutinating gelating-coated erythrocytes. Vox Sang. 12: 443–456 (1967).
261 Wong, G.; Pawelek, J.: Control of phenotypic expression of cultured melanoma cells by melanocyte stimulating hormones. Nature, Lond. 241: 213–215 (1973).
262 Yamada, K.M.; Olden, K.: Fibronectins – adhesive glycoproteins of cell surface and blood. Nature, Lond. 275: 179–184 (1978).
263 Yamada, K.M.; Schlesinger, D.H.; Kennedy, D.W.; Pastan, I.: Characterization of a major fibroblast cell surface glycoprotein. Biochemistry 16: 5552–5559 (1977).
264 Yamada, K.M.; Weston, J.A.: Isolation of a major cell surface glycoprotein from fibroblasts. Proc. natn. Acad. Sci. USA 71: 3492–3496 (1974).
265 Yucas, M.; Sugita, M.; Bensam, A.: A model of cell size regulation. J. theor. Biol. 9: 444–470 (1965).
266 Yeh, J.; Fisher, H.W.: A diffusible factor which sustains contact inhibition of replication. J. Cell Biol. 40: 382–388 (1969).
267 Young, C.W.; Hendler, F.J.; Karnofsky, D.A.: Synthesis of protein for DNA replication and cleavage events in the sand dollar embryo. Expl Cell Res. 58: 15–26 (1969).
268 Zetterberg, A.; Auer, G.: Proliferative activity and cytochemical properties of nuclear chromatim related to local cell density of epithelial cells. Expl Cell Res. 62: 262–270 (1970).

P.A. Riley, MB, PhD, Department of Biochemical Pathology,
University College School of Medicine, University Street,
London WC1E 6JJ (England)

# Subject Index

Adenocarcinomas, GI tract in organ
  culture 60
Adhesion
  cells to culture substrata 147–153
  role of fibronectin 147
Adrenocorticotrophic (ACTH) hormones,
  effect on cell proliferation 180
cAMP
  and cGMP in cell division 12
  mediation of hormonal effects 182–184
Apoptosis, development 11

Carcinoembryonic antigen, production by
  tumours in organ culture 74
Carcinomas, organ culture 61, 97–100
Cell division (proliferation)
  adhesion of cells 147–153
  blood flow 29–33
  cAMP/cGMP role 12
  cell density and rate 161–171
  cell flattening 155
  control 11–14, 131–198
    trauma 17
  cytoplasm/DNA ratio 9, 135
  diurnal variation 5
  effects
    hormones 23, 37
    in vitro resource depletion 158
    prostaglandins 33
  extrinsic factors in regulation 140–177
  hierarchical controls 133
  hydrogen ion concentration 176–178
  inducers of DNA synthesis 138–140
  intercellular control 140–158
  interpopulation control 178–184
  intrinsic factors in regulation 134–139
  long range intercellular effects 171–176
  metabolic production control 171–176
  proliferative control systems 132
  role of acute inflammation 28, 33
  role of vascular changes 29–32
  short range intercellular effects
    171–176
  spatial constraints in control 140–147
Cell movement, contact inhibition
  153–171
Cell turnover, cessation of growth 4
Chemotherapy, assessment of tumour
  responses in vitro 89–93
Compensatory growth
  lung alveolar tissue 25–28
  regulatory factors 25–28
Cortisol, effect on cell proliferation 180

Differentiation
  concept of neoplasia 2
  tumours induced by embryonic
    inductors 75–78
  tumours in organ culture 69–78

Embryonic inductors, tumour differentiation 75–78

## Subject Index

17 β-Estradiol, effect on
  cell proliferation 180
  mammary carcinoma and fibroadenoma 82
Estrogen, effects on endometrial carcinoma 79

Fibroadenomas, organ culture 62
Fibronectin
  mitosing cells 152
  nomenclature 151
  role in cell adhesion 147–153
  virus-transformed cells 152

Glucagon, effect on cell proliferation 180
cGMP
  and cAMP in cell division 12
  mediation of hormonal effects 182–184
Growth
  cell loss 8
  compensatory 5, 25
  control by nervous system 21
  developmental 4
  embryonic 3
  effects of hormones 23, 178–184
  factors for cells in culture 23, 180
  intracellular programming 8–11
  lung alveolar tissue 25–28
  lymphoid system in regulation 20, 36
  neoplastic 7
  neural mechanisms in control 32
  physiological load and regulation 19, 36
  regeneration 5
  regulation, through negative feedback 17–19, 35
  role of hormones 23, 37
  tissue interaction in control 24
  tumour organ culture, evaluation 65–69
  types 3–8
  wound healing 5
Growth hormone, production by somatotrophin adenomas in organ culture 73

Hormones
  effects on cell proliferation 178–182
  mode of action, cell proliferation 182–184

Hydrogen ion concentration
  intracellular, cell cycle 178
  metabolic process and cell proliferation 176–178
Hyperplasia, increase of functional capacity 7

Insulin, effect on cell proliferation 180
Invasiveness
  in vitro studies 83–89
  tumours 82–89
Ionising radiation, assessment of tumour responses in vitro 89–93

Lung alveolar tissue
  compensatory growth 25
  growth regulation 25–28

Melanotropin, effect on cell proliferation 180

Neural mechanisms, chemical mediation of growth control 32, 36

Organ culture
  antigenic expression of tissues 74
  characteristics of tumours 59–78
  differentiation and function 69–78
  evaluation of growth 65–69
  factors affecting tumour growth 55–58
  methodology 53–55
  production of carcinoembryonic antigens 74
  production of hormones 72
  response to hormones by tumours 78–82
  tumours 53–59, 97–105

Plasminogen activator, production by ovarian adenocarcinoma 73
Progesterone, effects on
  cell proliferation 180
  endometrial carcinoma 79
  mammary adenocarcinoma 80
  mammary development 80

## Subject Index

Prolactin
  effect
    cell proliferation 180
    lactalbumin production 81
    rat mammary carcinoma 81
  production by tumours in organ
    culture 73
Prostaglandins
  effects on cell division 33, 182
  mediation of inflammatory process 15

Somatomedin, effect on cell proliferation 180

Trauma
  mediation of inflammatory responses 14
  tissue response 14–16

Tumours
  assessment of response to chemotherapy,
    in vitro 89–93
  assessment of responses to ionizing
    radiation 89–93
  differentiation in organ culture 69–78,
    75–78
  effects of hormones on organ
    culture 78–82
  growth in organ culture 53–78, 97–115
  invasiveness 82–89
  longevity of organ cultures 59–65
  response to virus infection 93–95

Wound hormones 17, 35

THE LIBRARY
UNIVERSITY OF CALIFORNIA
San Francisco
666-2334

**THIS BOOK IS DUE ON THE LAST DATE STAMPED BELOW**

Books not returned on time are subject to fines according to the Library Lending Code. A renewal may be made on certain materials. For details consult Lending Code.

| 14 DAY | 14 DAY | |
|---|---|---|
| NOV 23 1985 | SEP -1 1989 | |
| RETURNED | RETURNED | |
| NOV 18 1985 | SEP -5 1989 | |
| 14 DAY | | |
| MAR 10 1986 | | |
| RETURNED | | |
| FEB 26 1986 | | |
| 14 DAY | | |
| MAY 9 1989 | | |
| RETURNED | | |
| MAY -9 1989 | | |

Series 4128